FROM OCCAM'S RAZOR TO THE ROOTS OF CONSCIOUSNESS

To

Elizabeth Rosenthal
a kind and scholarly friend

From Occam's Razor to the Roots of Consciousness

20 Essays on Philosophy, Philosophy of Science and Philosophy of Mind

GERHARD D. WASSERMANN
formerly Reader in the Theory and Philosophy of Biology
University of Newcastle upon Tyne

Avebury

Aldershot • Brookfield USA • Hong Kong • Singapore • Sydney

Published by
Avebury
Ashgate Publishing Limited
Gower House
Croft Road
Aldershot
Hants GU11 3HR
England

Ashgate Publishing Company
Old Post Road
Brookfield
Vermont 05036
USA

British Library Cataloguing in Publication Data

Wassermann, Gerhard D.
From Occam's razor to the roots of consciousness ; 20 essays on philosophy, philosophy of science and philosophy of mind. - (Avebury series in philosophy)
1. Philosophy
I. Title
101

Library of Congress Catalog Card Number: 96-86338

ISBN 1 85972 285 7

Typeset by
Kris Morgan
PO Box 250
Oxford, OX1 1AP

Printed in Great Britain by the Ipswich Book Company, Suffolk

Contents

Note about the author

The author, who was Reader in the Theory and Philosophy of Biology in the University of Newcastle upon Tyne, England, started his career as an applied mathematician and theoretical physicist with a first class Honours degree and a Ph.D. from the University of London, and then, for some time, studied biology at Birkbeck College (University of London). For many years he has specialized in various closely interrelated topics in theoretical biology as well as connected aspects of biophilosophy, philosophy of science and philosophy of mind. Apart from numerous publications in specialized journals he has published *Molecular Control of Cell Differentiation and Morphogenesis* (1972, Marcel Dekker, New York), *Brains and Reasoning* (1974, Macmillan, London), *Neurobiological Theory of Psychological Phenomena* (1978, Macmillan, London), *A Philosophy of Matter and Mind* (1994, Ashgate Publishing, Aldershot), *Keys to Life* (to appear, Ashgate Publishing, Aldershot), and *Shadow Matter and Psychic Phenomena* (1993, Mandrake of Oxford, Oxford). He was also an Associate Editor of the *Bulletin of Mathematical Biology*, a visiting Professor of the Hebrew University of Jerusalem and a visiting Professor of the Institute of Evolution of the University of Haifa (Director Professor Eviatar Nevo, Foreign Member of the Linnean Society, London).

Acknowledgements

Chapter 2 contains the whole of my article 'Theories, Systemic Models (SYMOs), Laws and Facts in the Sciences' which appeared in *Synthese* in 1989, vol. 79, pp. 489-515. This article appears here reprinted by kind permission of Kluwer Academic Publishers, Postbus 17, 3300 AA, Dordrecht, Netherlands.

Chapter 4 contains the whole of my article 'On the Laws of Nature' which appeared in *Synthese* in 1982, vol. 51, pp. 381-396. This article appears here reprinted by kind permission of Kluwer Academic Publishers (address above).

Chapter 5 contains the whole of my article 'Biological Aspects of the Philosophy of Mind' which appeared in *Metaphilosophy* in 1980, vol. 11, pp. 199-209. This article appears here reprinted by kind permission of Basil Blackwell Ltd, 108 Cowley Road, Oxford OX4 1JF.

Chapter 6 contains the whole of my article 'Reply to Popper's Attack on Epiphenomenalism' which appeared in *Mind* in 1979, vol. 88, pp. 572-575. This article appears here reprinted by kind permission of Oxford University Press, Oxford Journals, Walton Street, Oxford OX2 6DP.

Chapter 7 contains the whole of my article 'Thoughts on Popper' which appeared in *Nature* (London) in 1981, vol. 293, p. 6. This article appears here reprinted by kind permission of Macmillan Magazines Ltd, Porters South, Crinan Street, London N1 9XW. Copyright Macmillan Magazines Ltd, 1981.

Chapter 8 contains the whole of my article 'A Critique of some Aspects of Human Ethology' which appeared in *Behavioral and Brain Sciences* in 1980, vol. 3, pp. 630-631. This article appears here reprinted by kind permission of Cambridge University Press, Edinburgh Building, Shaftesbury Road, Cambridge CB2 2RU.

Chapter 9 contains the whole of my book review 'The Chief Abstractions of Biology' by W.M. Elsasser (North Holland and American Elsevier). My review appeared in *Journal of Neurological Sciences* in 1976, vol. 30, p. 216 (it was invited by the editor Lord Walton). This review appears here reprinted by kind permission of Elsevier Science B.V., P.O. Box 521, 1000 AM Amsterdam, Netherlands.

Chapter 10 contains the whole of my article 'Kindly Explained' which appeared in *Nature* (London) in 1982, vol. 296 p. 284. This article appears here reprinted by kind permission of Macmillan Magazines Ltd (address above).

Chapter 11 contains the whole of my book review 'The Biological Origin of Human Values' by E. Pugh (Routledge and Kegan Paul) which appeared in *Philosophia* in 1982, vol. 11, pp. 407-411. This review appears here reprinted by kind permission of Bar-Ilan University Press, Ramat Gan, 52900 Israel, and kind permission of the editor of *Philosophia*, Professor Asa Kasher.

Chapter 12 contains the whole of my article 'Human Behaviour and Biology' which appeared in *Dialectica* in 1983, vol. 37, pp. 169-184. This article appears here reprinted by kind permission of Societe Dialectica, P.O. Box 5907, CH-3001 Bern, Switzerland.

Chapter 13 contains the whole of my article 'Neuropsychology vis-à-vis Skinner's Behaviourist Psychology' which appeared in *Behavioral and Brain Sciences* in 1984, vol. 17, pp. 700-701. This article appears here reprinted by kind permission of Cambridge University Press, Edinburgh Building, Shaftesbury Road, Cambridge CB2 2RU.

Chapter 15 contains the whole of my article 'Morality and Determinism' which appeared in *Philosophy* in 1988, vol. 63, pp. 211-230. This article appears here reprinted by kind permission of Cambridge University Press (address above).

Chapter 16 contains the whole of my article 'Morality and Determinism: A Reply to Flew' which appeared in *Philosophy* in 1990, vol. 65, pp. 99-101. This article appears here reprinted by kind permission of Cambridge University Press (address above).

Chapter 17 contains the whole of my article 'Wittgenstein on Jews: Some Counter-examples' which appeared in *Philosophy* in 1990. This article appears here reprinted by kind permission of Cambridge University Press (address above).

Chapter 20 contains the whole of my article 'Some Comments on Methods and Statements in Parapsychology and other Sciences' which appeared in *British Journal for the Philosophy of Science* in 1955, vol. 6, pp. 122-140. This article appears here reprinted by kind permission of Oxford University Press, Oxford Journals, Walton Street, Oxford OX2 6DP.

1 Introduction

This book deals with a selectd number of important topics in philosophy, philosophy of science and philosophy of mind. Most, but not all, of its chapters have been previously published as separate papers in various prestigious journals in the fields just mentioned. The present book, apart from adding some important new ideas, notably in Chapters 3 and 21, which are newly written for this book, brings together numerous ideas which I have developed over a period of more than 20 years. There is a small measure of overlap between the present work and my previous monograph *A Philosophy of Matter and Mind*, notably in Chapter 2. However, repetition of some of my earlier ideas which, in fact, in the paper presented in Chapter 2 preceded the monograph just mentioned, is important. In order to grasp some of the essentials of Chapter 3, namely why, in my opinion, Occam's celebrated 'razor' is useless when applied to the most prominent hypothetico-deductive theories in the sciences, it is important to gain some idea of the essentials of the structure of typical hypothetico-deductive theories in the sciences. In fact, I have only relatively recently, in 1989 in the journal *Synthese*, succeeded in expounding the characteristic structure of various, well known, hypothetico-deductive theories, such as Newtonian Mechanics, Classical Electromagnetic theory, Quantum Mechanics etc. This may help to explain why until now it was not obvious that there is something very wrong with Occam's razor, which invalidates that often cited instrument. Many scientists and philosophers have been told that it is a virtue, when theorizing, to be parsimonious in the invocation of hypotheses, and that good theories are those that use very few hypotheses. This, however, is simply not true, as will be argued in Chapter 3. Typically, Newtonian Mechanics, and other celebrated theories, depend each, when applied to various fields of science and engineering, on hundreds, and, more likely, several thousands of distinct types of hypotheses. That this is so becomes apparent with the help of my analysis of typical scientific hypothetico-deductive theories in Chapter 2.

Apart from laying the foundations for Chapter 3, one of the most interesting results of Chapter 2 is the demonstration that most or all scientific hypothetico-deductive theories have the same characteristic structure (see also Wassermann, 1994). This cannot be entirely a matter of the human brain making its owner think so as to impose the same characteristic structure on theories dealing with natural phenomena, say of mechanical systems. The sameness of the characteristic structures under discussion must also have something to do with the physical systems dealt with by the hypothetico-deductive-theories. At present various theoretical physicists are trying to discover unified theories that embrace all known major hypothetico-deductive theories of physics (with certain exceptions, such as classical mechanics). Such aimed at theories are referred to as *Theories of Everything* (or TOEs for short). If any suitable TOE should be discovered then, perhaps, deeper reasons might emerge why the most prominent hypothetico-deductive theories of theoretical physics have the same characteristic structure.

What, actually, had led me to say farewell to Occam's razor? Above all it was the realization that numerous thinkers from village idiots to Nobel laureates had paid homage to it, while in practice little, if any, use was being made of it. I had been urged, in my own theorizing by well-wishers and foes to be economical in the use of hypotheses, because this was the 'done thing'.

After leaving behind Occam's razor in Chapter 3, I shall turn to another topic which has involved problematics. Many philosophers have used statements of the kind 'All *X* are *Y*'. Such statements, as I shall show in Chapter 4, lead to Hempel's celebrated paradox. Many people have tried to resolve that paradox, but unsuccessfully. My own attempt, which may be successful, is, as various people told me, similar to Kyburg's. I shall argue, in Chapter 4, that statements of the kinds 'All *X* are *Y*' are wrongly formulated and that, instead, they ought to be replaced by statistically formulated statements. When this is done, Hempel's paradox seems to vanish. That everyday empirical generalizations ought to be formulated statistically is, perhaps less obvious than the fact that many empirical generalizations of physics (e.g. Ohm's law, or Boyle's law) are concealed statistical statements. It is often thought that only Quantum Mechanics has a thoroughly statistical character, whereas this does no apply to most other hypothetico-deductive theories of physics. Indeed, when the statistical nature of Quantum Mechanics was first recognized by Nobel Laureate Max Born, there were grave doubts by some eminent physicists. Einstein reputedly proclaimed that 'God does not play dice'. To the way of thinking of many scientists, and others, at that time, it was even less obvious that, say, Ohm's law should be of a statistical kind. That the resolution of Hempel's paradox should bring these issues to light shows that quite innocent philosophical arguments can illuminate profound issues in the sciences, at least sometimes.

The philosophical clarifications of the nature of Quantum Mechanics, particularly the statistical interpretation of Quantum Mechanics, just mentioned, shows clearly that science and philosophy cannot be rigorously separated. There are, however, numerous philosophers, but not philosophers of science, who believe that philosophy and science cannot be regarded as interlinked. Their thesis derives, perhaps, from Wittgenstein's philosophy and from the views of other philosophers who thought that philosophy should be free of hypotheses. Since science depends, through and through, on hypotheses (and deductions from hypotheses) these philosophers jumped to the conclusion that science and philosophy must not be mixed. The conclusions of Chapter 4 show already, by a possible resolution of Hempel's paradox, that the demanded separation of science and philosophy is utterly unrealistic. In fact, in Chapter 5 it is argued that the Philosophy of Mind has important biological aspects, showing further examples of the interrelation of science and philosophy. I have attended not a few philosophical seminars where some philosophers became quite livid when anyone tried to relate science to philosophy. In those days philosophy of science was still largely unknown in various British universities. Perhaps those days have gone, but when I published the paper in 1980 which is here reproduced in Chapter 5 the urge to separate science and philosophy was still very prominent, perhaps due to the fact that most philosophers then were not philosophers of science.

In Chapter 6 I am turning to the Philosophy of Mind. It is a short paper, but it exposed a fundamental error in the late Sir Karl Popper's thinking about epiphenomenalism. He thought that he could invalidate epiphenomenalism by means of evolutionary arguments. Not surprisingly, this turned out to be an unsound adventure. Popper, of course, was an interaction dualist. Epiphenomenalism forms an ingredient of some (but not all) brands of materialism, and is, thus, a thorn in the eyes of interaction dualists like Popper. This, however, is no reason to use spurious arguments to discredit epiphenomenalism.

Next, in Chapter 7 I have turned, once more, to Popper, and presented a very short note which appeared in *Nature* (London) in September 1981. Although I have drawn attention to Popper's mistaken views on the nature of the theory of evolution in two extensive papers, which are reproduced in a forthcoming monograph (titled *Keys to Life, Philosophy and New Mechanisms of Evolution and Development*, to be published by Avebury, Ashgate Publishing, 1997), Popper in his own recantation of his views in his Darwin Lecture (Darwin College, Cambridge, 1977) provided again inappropriate arguments which were far off the mark. Indeed, I (and probably others) were surprised that with so many illustrious biologists about in the field of evolution that Popper was asked at that time to give the Darwin Lecture. I am not denying that Popper was a very great man, but his arguments in the field of biology were repeatedly erroneous.

Chapter 8 contains an invited *Peer Comment* in the journal *Behavioral and Brain Sciences*, titled 'A Critique of some Aspects of Human Ethology' which appeared in 1980. Ethology is, perhaps, primarily concerned with 'innate' (i.e. largely genetically controlled) behaviour. Yet some forms of behaviour also depend on environmental control. Moreover, granted that some aspects of such behaviour are genetically controlled, it remains to isolate the genes for such behaviour. Eibl-Eibesfeldt, who is a distinguished pioneer in the field of human ethology, provides many intriguing questions and possible answers for those interested in his subject. I must confess that for me there are more questions than satisfactory answers. There is a temptation to compare human innate behaviour with animal innate behaviour which is highly stereotyped, whereas human 'innate' behaviour depends strongly on intricate cognitive capacities of a kind that are not largely genetically controlled. It is argued that genes code for nervous system capacities in man and not for the actual use made of these capacities. Some of the views of Eibl-Eibesfeld may raise serious doubts in sceptics of the validity of the whole field of human ethology. Only time can tell how far studies of this subject can succeed. My discussion of Eibl-Eibesfeldt's position and of his subject also refer to numerous topics which I shall not refer to at this stage, and which readers are advised to examine in detail.

After my discussion of human ethology I have reproduced in Chapter 9 my review of W.M. Elsasser's book *The Chief Abstractions of Biology*. In 1976 Professor (now) Lord Walton at Oxford University, then Dean of the faculty of Medicine at the University of Newcastle upon Tyne, invited me to review Elsasser's book for the *Journal of the Neurological Sciences* which Lord Walton edited. Elsasser was renowned as a leading theoretical geophysicist. Yet his contribution to theoretical biology and biophilosophy in this book may not be to everybody's taste. Elsasser seems to over-emphasize the individual variability of organisms, which is partly due to different organisms often inheriting different variants (alleles) of genes. Yet, despite this variability the basic mechanisms of most (but not all) organisms of the same species are similar, and Elsasser's underlying anti-mechanistic trend seems to me uncalled for. I have stressed that, in my opinion, Elsasser's abstractions lack explanatory value, and obscure the essentials of mechanisms. What is perhaps surprising is that a great variety of dogs, comprising, among others, spaniels, dachshunds, fox terriers, poodles and sheep dogs, have, despite striking variations in sizes, shapes and coloration, retained similar biological machinery for sustaining their lives. It is only by elucidating the details of this machinery that we can understand how these organisms work. I do not think that the philosophy of holism, which regards organisms as 'wholes' and of 'organicism', both of which Elsasser advocates, can get us far, if anywhere, in understanding the detailed mechanisms of organisms.

In Chapter 10 I am giving a brief discussion of creationism, which was published by the journal *Nature* in 1982. Readers who want to study my much more extensive discussion of creationism might wish to turn to my new book *Keys to Life,* mentioned above, which devotes much of a chapter to a critique of creationism. In my 1982 note I stress that there are many different types of explanations of the facts that evolutionists deal with. Some people, like myself, will opt for mechanistic materialistic explanations. Others may prefer creationist interpretations of the facts. What people opt for in these circumstances may depend on their upbringing and education and contact with particularly biased people. Perhaps few fields of human endeavour reveal such strong biases as the topic of evolution. Some religious believers maintain adamantly that there is something drastically wrong with the neo-Darwinian theory of evolution and that God created the Earth and all the fossils in it. Some mechanists, on the other hand, hold diametrically opposed views. I think that it is futile to argue in serious ways with such strongly opposed views. One can only assert where one's own sympathies lie. Many creationists have doubts about evolution because in their opinion neo-Darwinian theory relies too much on random events, and for many people random events seem no way for evolving the intricate structures and innate behaviour of organisms. For my part I have argued in my book *Keys to Life* (to appear in 1997, chapters 4, 5 and 6) that my postulated TIMA mechanisms can dispense with much randomness, but retain some, and yet explain the evolution of much that is characteristic of complex organisms.

On turning from creationism to my book review of George Edgin Pugh's book *The Biological Origins of Human Values*, published in 1978 and reviewed in 1982, one soon gets the impression - rightly or wrongly - that Pugh is not one of those people whose work excites everybody. Unlike animal sociobiology human sociobiology (like human ethology) does not set the Thames on fire. Pugh is certainly a brave man to venture into this controversial field. My review of Pugh's book may seem harsh to some sociobiologists. But then, my original studies in science were in the 'exact' sciences, notably theoretical physics, where people like precision of thinking. I have mainly included this short book review so that readers who have not already sampled this field can get an idea of the deep division that separates theories in the exact sciences, discussed in chapter 2, from theories in fields such as sociobiology.

I must now turn to a brief mention of Chapter 12, which is concerned with *Human Behaviour and Biology*. The abstract of the paper reads as follows:

> Extremism in the environment-versus - innateness controversy in the behavioural sciences and in human sociobiology is being examined. Genetic

effects can be severely modified or overruled by environmental factors, but may, nevertheless, be important. Dawkins' view that we are survival machines, programmed to subserve selfish genes seems untenable and is a root of racialism. It is also argued that morality is compatible with mixed genetic and environmental control of brains via existing biological machinery.

Chapter 13 deals with Skinner's behaviourist psychology which now, at a time when cognitive psychology has once more come to the fore, looks perhaps, like a psychological dinosaur. Nevertheless, my commentary was invited as a *Peer Comment* by the journal *Behavioral and Brain Sciences.* This Peer Comment was also reissued in Skinner's collection of canonical papers by Cambridge University Press in 1988. My main concern in that paper was to show that Skinner paid little, if any, attention to neuropsychology. This is undesirable, since human behaviour is intimately related to neural systems.

In Chapter 14 I have 'argued that "rationality" is something distinct from intelligence and goal-directed thinking. I have also criticized attempts to formalize theoretical and practical rationality, and have argued that a satisfactory formal definition of "rationality" seems unobtainable, mainly because of the difference between relativists and anti-relativists. It is suggested that any alleged "rationality" of people is not factual but hypothetical, and, like all hypotheses, hypothesized rationality cannot be proved to be true. Its alleged "truth" can at best be a matter of consensus. The kinds of so-called "rationality" present among similarly reasoning people depend on educational contingencies as well as on the constitution of the brains of such people. Neuropsychology may help to explain how similarly constructed brains could reach consensus. It is argued that consensus among a large social group, or often a small body of experts, or a body of believers in the case of a religion, is essential if some mode of reasoning, action or decision is to be classified as "rational" in the sense of being "acceptably meaningful and plausible" to the group(s) concerned. Popper's views on critical rationality are rejected in favour of pragmatism, and Newton-Smith's anti-relativistic arguments are briefly criticized. Sperber's critique of anthropological research in relation to the relativity of rationality, although interesting, leaves the central contention of relativism intact.' (Cited summary of 'Social Aspects of Rationality' in *Methodology and Science* (1984), *17*, 50-69.)

In Chapter 15 I shall reproduce my paper *Morality and Determinism* first published in *Philosophy* in 1990. As I wrote in the introduction to the paper:

> This paper is intended as a contribution to a recent vigorous debate in *The Times*, between the distinguished journalist Bernard Levin, the eminent

Oxford economist Wilfred Beckerman and the [then] Archbishop of York, John Habgood, among others.

About halfway through the paper after a lengthy debate of whether human evil, where it exists in people, is innate, I wrote that:

> I conclude at this stage that, although many psychopathic personalities *could* have a genetic basis for their personality defect, no reliable information exists (except for certain schizophrenics) concerning any possible genetics of such disorders, and such disorders are also difficult to diagnose. I agree with Bernard Levin that, often, only a *handful* of psychopaths can start, and keep going, massive onslaughts on people's lives and/or their property, and that this can happen in concentration camps, labour camps, in rioting crowds or among football hooligans.

Also in the same paper 'I conclude that Beckerman's main argument (cited at length in Chapter 15) is unsatisfactory and that *any metaphysical belief in "determinism" does not force us to consider the passing of moral judgements "an absurdity"*.' The detailed reasoning that led me to this conclusion is presented in Chapter 15.

Not long after publication of 'Morality and Determinism', the philosopher Flew published a short note criticizing my paper. I rejected his criticisms as unsound in a short note in *Philosophy*, which is published here as chapter 16. I opened that note with the statements:

> Antony Flew's criticisms of my paper 'Morality and Determinism' seem to me flawed because of (1) Flew's disregard of the existence of extensive statistical (probabilistic) determinism in the natural sciences and (2) Flew's out-of-context quotations of parts of my sentences with resulting misrepresentations of my meanings.

In Chapter 17 I have dealt with Wittgenstein's deplorable antisemetic allegations, by reproducing my paper 'Wittgenstein on Jews: Some Counter-examples', which I published in *Philosophy* (1990), *65*, 355-365. Wittgenstein was three quarters Jewish and considered himself a Jew. At the opening of my discussion note in *Philosophy* I stated that:

> Wittgenstein's empirical generalizations about the personality structure of Jews, in *Culture and Value*, as cited and reconstructed by Yuval Lurie, seem to me to be formally and factually of the same misguided genre as many untrue empirical generalizations about Jews in Richard Wagner's essay *Judaism in Music*, in the (forged) *Protocols of the Elders of Zion*, in Hitler's *Mein Kampf*, in the arch-anti-Semitic work by the late Nazi Nobel Laureate

Philipp von Lenard, *Deutsche Physik* (which claimed that one must distinguish Jewish from Aryan Physics), and in myriads of other anti-Semitic tracts produced before, during, or after the Nazi period.

I shall refute many of Wittgenstein's detailed allegations against the Jewish people, notably his invalid arguments concerning originality and genius of many Jews, by listing some of the many celebrated Jews of genius and originality who contributed to various sciences, the arts, literature, music and so forth. Henceforth some of Wittgenstein's admirers must, perhaps, conclude that , like the equally (or perhaps more) antisemetic composer Richard Wagner, Wittgenstein was an evil genius.

Chapter 18 reproduces my paper 'Moderate Relativism and Psychological Sets' first published in *Methodology and Science* (1987), *20*, 223-237. At the start of that paper I wrote that the paper 'aims to propound and defend a variety of moderate relativism (MR) which avoids extreme relativist claims and which partly relies on the notion of "psychological set"(PS).' The summary of the paper reads:

> A moderate relativism (MR) is expounded and defended. It is argued that some of our 'psychological sets' (PSs) are partly or entirely determined by the views of social groups to which we belong either overtly or by tacit agreement. PSs are apt to generate strong prejudices in favour of certain belief systems and to overrule considerations of alternative views. Popper's 'critical rationality' is suggested to be utopian since it underrates the strength of PSs. The views of Sperber, Lukes and other partial, or strong, anti-relativists are examined and criticized. It is also stressed that even the *notion* of truth is relative to certain group-accepted 'theories of truth'. Some 'universalist' claims (e.g. those of Gellner) seem to be based on over-optimism and the claims of the present MR seem to be closer to practically encountered situations than these universalist claims.

After this excursion into social philosophy, I have returned in Chapter 19 to the paper 'Quantum Mechanics and Consciousness', which was published in *Nature and System* (1983), *5*, 3-16. The paper was also refereed by the more prominent journal *Erkenntinis*, who informed me that they would have accepted it. But by that time I had already given it to *Nature and System* to publish. To give readers an idea what the paper is about, let me cite the beginning of the paper, which reads:

> In this paper I shall discuss one common but widely accepted fallacy and another less familiar fallacy concerning the relation between human consciousness and quantum mechanics. The first fallacy was propounded in a book review by Marie Jehoda in *Nature* (London). She writes: 'The

revolution in the basic assumptions of physics brought about by Einstein and by the development of quantum theory has repercussions far beyond that discipline. One of the consequences of the new physics - that the mind of the observer is a necessary ingredient of the structure of theory - has not only returned terms like "mind" or "consciousness" to respectable scientific status, more far reaching in the views of growing numbers of scientists, philosophers and historians of science it also undermines the mechanistic world view that originated with Bacon, Descartes and Newton.'

I shall argue (in Chapter 19) that these conclusions simply do not follow if one embraces a defensible kind of materialism, such as epiphenomenalism or the closely related 'materialistic dualism' which I have recently espoused (cf. *Review of Metaphysics* (1982), *35*, 715-729) and which some people regard as being close to a dual aspect theory.

Another apparently fallacious argument that will be dealt with in Chapter 19 is the great physicist Wigner's contention that, to cite Landsberg:

> Quantum mechanics leads to practically zero probability for the existence of self-reproducing states. He [Wigner] concluded that either present quantum mechanics cannot furnish an explanation of all that exists, or it should be modified to include aspects, such as consciousness, which are needed when dealing with purely physical problems.

In Chapter 19 Wigner's argument will be dealt with and disposed of.

Next, in Chapter 20 I am reproducing a paper originally published in *British Journal for the Philosophy of Science* (1955) *6*, 122-140. This paper arose out of discussions with (now) Professor Ronald Harrop of Simon Fraser University Canada. One of the central topics of the paper concerned repeatability of scientific observations. There exist, as will be shown, two kinds of repeatability, and one of these concerns what *was* repeatable in past observations, and does not imply future repeatability. The other type of repeatability refers to forecasts of future repeatability on the basis of past repeatability. This distinction is important. For instance, in parapsychology one deals in most reports with statements concerning past repeatability but not with predictions of future repeatability (which often does not occur). The discussions of Chapter 20 are also relevant to sciences other than parapsychology.

The confusion of the two kinds of repeatabilities is very common, notably among parapsychologists past and present. Indeed, when people call for the need of repeatability in parapsychology then this does not necessarily imply indefinite future repeatability of a state of affairs, but may not require more than that a state of affairs was observed repeatedly in the past under stated conditions. In fact the repeatability may not relate to more than *class characteristics* of members of a certain class of phenomena, rather than repeatability of 'data'. For instance, anatomists may claim that they have observed repeatably certain types of structures with specifiable class characteristics, when studying human bodies. From this we cannot infer that for all eternity such structures will repeatedly occur in precise detail. Our knowledge of evolution certainly suggests that such a forward projection is unwarranted. In evolutionary studies we can only state with confidence what *was* observed repeatedly, but not what will be observed repeatedly, say over a long span of thousands of millions of years.

Finally, in Chapter 21 I shall deal with the nature of consciousness. Consciousness is a topic that fascinates laymen, scientists and philosophers alike and many hundreds, or, more likely, thousands, of papers and also many books have been dedicated to that subject. In this situation one cannot, short of writing a book on the subject, do justice to the nature of consciousness. Faced with this situation, I have fallen back on a paper by Hebb published in 1954 in a symposium titled *Brain Mechanisms and Consciousness* (publishers, Oxford, Blackwell Scientific Publications). I suppose that many people will argue that after more than 40 years Hebb's views might be dated. As regards the minutiae of his article this may well be so, but, since the nature of consciousness remains as much of an enigma as in 1954, one can fall back on those early days up to a point. Admittedly one knows more now about the biological (neuropsychological) correlates of consciousness, but not by a long way enough to give an assured answer when questioned about the nature of consciousness.

This concludes the preliminary survey of what this book contains. Among its numerous topics the ones that have occupied most of my attention in recent years are those of chapters 2 and 3. I realized long ago that there exists, apparently, no comprehensive analysis of the nature of scientific hypothetico-deductive theories, which play such a central part in science. So I made it my task to provide such an analysis in the paper which is now reproduced in Chapter 2. Quite recently somebody not well versed in philosophy of science approached me. He wanted to tell me that one of my recently published theories contained too many hypotheses. Quite sarcastically the fellow wrote to me 'Surely, Dr. Wassermann, you must have heard of the Principle of Parsimony of hypotheses in science'. He meant Occam's razor. As a result of such insolent sarcasm, I decided to write down some ideas that I had had for some time about Occam's razor, namely that it simply does not apply to most or all scientific theories, for reasons stated in Chapter 3. In the life and times of William of Occam his 'razor' would have attracted theologians and philosophers, among others. Occam's razor could give apparent answers to very naive metaphysical questions. For instance, fewer hypotheses were, apparently, required to postulate that there was *only one* God, while many more hypotheses were needed in order to postulate many deities (as, e.g. in ancient Greek theology). In such situations it could be held that Occam's razor permitted to decide which belief system was the better. But even this is quite doubtful.

Reference

Wassermann, G.D. (1994) *A Philosophy of Matter and Mind* (Ashgate Publishing Ltd., Aldershot)

2 Theories, systemic models (SYMOs), laws and facts in the sciences

2.1 Aims of the paper

The gist of this paper is constructive and provides a partly novel characterization of the structure and function of scientific hypothetico-deductive theories, and also provides applications of some of my ideas to evolutionary theory. Apart from this the paper examines whether certain kinds of scientific theories can generate and/or strengthen a belief in the 'reality' and 'existence' of certain systems. I propose to show that:

(1) A *scientific hypothetico-deductive theory* (SHDT) comprises an open set of *scientific systemic models* (SYMOs);

(2) Each SYMO of a particular scientific hypothetico-deductive theory is based on a set of '*general starting hypotheses*' (GSHs) of that SHDT, typical examples of 'general starting hypotheses' of SHDTs are Newton's three laws of motion of classical mechanics, or Maxwell's equations of Electromagnetic Theory;

(3) Apart from using 'general starting hypotheses' each SYMO depends on 'SYMO-specific *ad hoc* hypotheses', which usually idealize (such as the assumed 'light', i.e. massless, string of a simple pendulum, and the assumption that the bob of the pendulum is a point particle);

(4) In general a SYMO serves to deduce a number of '*empirical generalizations*' (sometimes called '*empirical laws*' after their discoverers, such as Ohm's law, Boyle's law, Kepler's laws, etc.);

(5) We can formally characterize the *j*th SYMO, SYMO_{ji}, of the *i*th SHDT, SHDT_i by

$$(1)\ (\text{GSHs of SHDT}_i) + (\text{SSHs of SYMO}_{ji}) \rightarrow \{\text{EG(d)}_{ji}\}$$

where SSHs denote the 'SYMO-specific *ad hoc* hypotheses' and $\{EG(d)_{ji}\}$ denotes the set of all empirical generalizations (or empirical laws) that are, known to be, deducible for $SYMO_{ji}$ by combining the SSHs with the GSHs of $SHDT_i$. It must be stressed that $\{EG(d)_{ji}\}$ represents an *open* set at any time of the history of $SYMO_{ji}$, since further elaboration of $SYMO_{ji}$ (eg. modifications of its SSHs, or successes in making further deductions) could lead to additional EG(*d*)s of set $\{EG(d)_{ji}\}$. An EG(*d*) will, in the sequel, be referred to as an '*empirical generalization*$_{deduced}$'.

(6) One major thesis of this paper is that empirical data can be used to test and confirm (or, respectively, falsify) the claims of validity of a SYMO, *without* thereby testing or confirming the *existence* and *reality* of the system modelled by the SYMO. An SYMO *can be tested by testing its* 'empirical generalizations$_{deduced}$'. It cannot be overstressed, as can be seen at a glance from scheme (1) above, that confirmation of an 'empirical generalization$_{deduced}$', of set $\{EG(d)_{ji}\}$ in the scheme above, only confirms the temporary acceptability of the *combined* use of GSHs and SSHs of $SYMO_{ji}$, but it does *not* confirm the individual validity of each GSH and SSH used. Accordingly, what observations and experiments confirm about a SYMO are only its 'empirical generalizations$_{deduced}$', but observations and experiments are, as a rule, not intended to confirm the validity of particular idealizing 'SYMO-specific *ad hoc* hypotheses'. Confirming the deductive results of a *combination* of hypotheses does *not* necessarily confirm the validity of either any individual 'general starting hypothesis' (and its idealizations) or any of the idealizing 'SYMO-specific *ad hoc* hypotheses' that constitute the combination of hypotheses used in a SYMO. Nevertheless, it happens not infrequently that 'improved' SYMOs are obtained by replacing a particular idealizing 'SYMO-specific *ad hoc* hypothesis' by a related, but *less* idealizing, 'SYMO-specific *ad hoc* hypothesis', leading thereby to better agreement of the modified SYMO with observation and/or experiment. Yet, even this tells us only that the new *combination* of hypotheses is 'better' than the preceding one, without thereby directly confirming the general validity of the newly invoked 'SYMO-specific *ad hoc* hypothesis'.

To facilitate reading of this paper I shall, apart from the acronym SYMO, avoid all acronyms used above, and write technical terms out in full.

Several asides in this paper deal with A. Olding's onslaught on some of the views expressed by Sir Karl Popper, Professor R. Lewontin (Harvard) and myself. Although a part of this paper serves to dismiss Olding's views, most of this paper has a positive mission and should be of interest to a wide circle of philosophers of science, and not just to bio-philosophers. It should be of particular interest to philosophers who wish to understand the basic structure of scientific hypothetico-deductive theories and SYMOs developed within these theories.

2.2 Introduction and the comparison of SYMOs and 'really existing' systems

The relationship between a scientific hypothetico-deductive theory and the facts which that theory is intended to *explain* is notoriously intricate. Hence, one major aim of this paper is to analyze the relation between scientific hypothetico-deductive theories (or other kinds of scientific theories) and the facts. Although familiar to many experimental and theoretical scientists, usually through praxis, rather than explicit analysis, the relation between scientific hypothetico-deductive theories and facts may be less familiar to some philosophers. This became clear to me on reading a polemical paper by Olding (1984) which was directed mainly against positions held by Popper, Lewontin, and myself. Unlike Popper and many others, Olding does not seem to appreciate that scientific theories do not reveal 'absolute truths' or 'ultimate realities'. Instead, as I shall argue, scientific theories, in particular scientific hypothetico-deductive theories, depend on *scientific systemic models* (SYMOs) which *idealize* about real systems.

(Prompted by a referee, I shall slightly sidetrack at this point. Some philosophers, notably Popper, who realize that one cannot prove the truth of scientific theories, have considered, instead, the notion of '*verisimilitude*'. As Oddie (1981) wrote in his review of that topic, 'The problem of verisimilitude (truthlikeness) is to specify rigorously what it takes, in general, for one theory to be *closer to the truth* (or *more truthlike*) than another'. Hitherto proposed measures of verisimilitude have only been applied to very artificial types of models and not to SYMOs, and the topic remains problematic. One would expect that any tentatively accepted numerical value of the measure of verisimilitude adopted for a scientific theory, vis-à-vis competitor theories, might need repeated upward and/or downward revisions in the light of new discoveries. It is not clear what such formal numbers, allocated to verisimilitude measures, could add to the enlightened commonsense evaluations of scientific theories by practicing scientists. Does it really help to be told that theory A is 3.287 units of verisimilitude above competitor theory B)?

At any rate, SYMOs are not meant to reveal 'absolute truths' about systems or to represent 'ultimate realities' of systems, and verisimilitude measures seem, in my opinion, far less useful than investigations of the actual and formal structure of scientific theories. SYMOs are not intended to explain in which sense 'real' systems actually 'exist'. Instead, SYMOs may be regarded as *brain-generated models* which try to represent in idealized forms some of the properties of presumed 'real' existing, systems. Some of the properties of a SYMO may only approximate to the properties of a 'real' system which the SYMO models. What many observations and experiments test and confirm (or falsify) are deductions from SYMOs, namely 'empirical generalizations $_{\text{deduced}}$' which are deduced from SYMOs. Since SYMOs are only meant to *model* properties of real systems, they are not iron-clad guarantors of the existence of real systems. Similar remarks apply to all kinds of other (non-SYMO, e.g. perceptual) brain representations of properties of supposedly real things. If, via a telescope, we observe light emitted by a distant star, then this is no guarantee that this star still exists. Again, that our brains somehow represent grass as green does not imply that there exists green-coloured grass or that its chlorophyll molecules are green. There are, we believe, molecules of grass whose emitted (or reflected) light can induce, via a sequence of complex physical and physiological processes, those brain representations in the grass-perceiver that make him perceive grass as green. Thus, the way things are modelled by brains need not correspond to the way things really are (or 'exist').

Often a SYMO, if its validity is confirmed by numerous tests of its 'empirical generalizations$_{\text{deduced}}$', may strengthen our belief in the postulated existence of certain real systems. Scientists have postulated many systems, which are sometimes referred to as '*hypothetical constructs*' and which cannot be directly observed as entities, by, say, visual perception. Typical hypothetical constructs are electrons, photons, quarks, W bosons, gravitons, gluons, neutrinos and positrons, among others. *One must not confuse hypothetical constructs and the* SYMOs *that model supposed theoretical properties of these hypothetical constructs*. Empirical data which confirm the validity of an 'empirical generalization$_{\text{deduced}}$' which is deduced from a SYMO used for modelling properties of a hypothetical construct do *not* thereby *confirm* the *existence* of real entities that correspond to the hypothetical construct, although *confirmations* of deductions from the SYMO may generate and/or strengthen our (metaphysical) *belief* in such real entities. Thus, Dirac's relativistic equation for the electron is a brain-generated SYMO. *Confirmation* of various of its 'empirical generalizations$_{\text{deduced}}$' may lead us to *believe* that there 'exist' real electrons. Yet, apart from generating *belief* in real electrons, Dirac's equation does not tell us in which sense or whether 'real' electrons 'exist' (see also Section 2.4 for various different SYMOs that have been used in the past to 'model' properties of electrons, although none of these models tell us what *real* electrons are like or in which sense they exist, e.g. as particles or superstrings). Even looking at the sun and perceiving a bright disc does not tell us in which sense the sun *exists in 'reality'* any more than perceiving green grass tells us in which sense grass exists in reality.

Nevertheless, when scientists talk about 'systems', such as electrons, quarks, photons, the sun, or grass and claim that these systems (some of which are hypothetical constructs) *'exist' and have particular properties,* then this is no more than a convention current in all science, and does not entail (logically) actual existence of the systems that are claimed to exist. Similar conventions occur in large numbers in everyday speech, e.g. when people say that grass is green (see above). For instance, if physicists say that 'electrons have spin ½', then this is a convention of speech widely adopted by many specialist scientists. What these scientists really mean is that a particular SYMO for the hypothetical construct 'electron', namely Dirac's equation, predicts as an 'empirical generalization$_{deduced}$' that, for instance, a hydrogen atom can give rise to particular (electron spin ½-related) spectral lines, which can be tested and have been confirmed to general satisfaction. Thus, *what is demonstrated is that a* SYMO *(i.e. Dirac's equation) has specific testable properties, and this has been confirmed by observations and tests.* What has *not* been observed or confirmed is that 'real' electrons with spin ½ exist . The verbal convention of talking about 'electrons having spin ½' might tempt some philosopher-logicians to conclude that 'real' electrons have been proved to 'exist'. These philosophers might argue (as did one referee of this paper) that if an object *b* (a 'real' electron) has property *F* (e.g. spin ½) then this *entails* that *b* exists. Yet, the preceding discussion shows that property *F* (in this case spin ½) has not been shown to be a property of a real object, *but has been shown to be a property of a* SYMO. But a property of a SYMO does not (logically) entail anything about the 'existence' of an object or system whose properties the SYMO claims to model. The analogy with green grass is again useful. Just as confirming deductions from a SYMO for an electron does not confirm in which sense (or that) real electrons exist, so the perceived appearance of green grass (via a brain representation induced by viewed grass) does not tell us anything about the sense in which real grass 'exists'. Thus, brain representations, whether SYMOs or perceptual (etc) brain representations, may serve to *model* presumed aspects of reality. But *confirmation* of SYMOs, or the perception of things or systems, do not entail the existence of systems, and *properties of* SYMOs, *or of other brain representations, are not necessarily properties of real systems that are modelled by* SYMOs *and other kinds of brain representations.*

I conclude that *pace* Olding (1984, p. 112), I am *not* denying 'that it is reasonable to assert the existence of anything' but I only claim that such assertions are conventions based on untestable *beliefs* in *unknowable* modes of real existence.

Statements that assert only a *belief* in the existence of systems or things (etc.), as distinct from assumed properties of SYMOs modelling these systems (etc.), were termed *'strictly existential statements'* by Popper (cf. 1983, p. 69), who stressed that such statements cannot be falsified by observed events. Typical 'strictly existential statements' are of the forms 'there exists an *X*' or 'there exists some *X*', where the 'strictly existential statement' only asserts the existence of an *X* but not any of the hypothesized properties of that *X*. Popper (1983, p. 69) noted that only a universal statement could contradict a 'strictly existential statement'. For instance, the strictly existential statement 'there exists an *X*' would be contradicted by the *universal* hypothesis 'there exists no *X*'. The latter hypothesis, however, cannot be shown to be true, since, as Lewontin (1972) noted, failure to find an *X* does not prove its non-existence, but might only mean that one has not looked hard enough.

I noted above the convention-based temptation (or habit) of many scientists to extrapolate from, say, well confirmed SYMOs to the 'existence' of the systems modelled by these SYMOs. This conventional extrapolation can lead to a confusion of SYMOs and 'strictly existential statements', as the following passage illustrates. I have slightly changed Popper's (1983, p. 69) text to suit present nomenclature:

> ...it might be asserted (with justice) that there are theories even in physics which have the form of 'strictly existential statements'. An example would be a 'strictly existential statement' deducible from the periodic system of elements, which asserts the existence of elements of certain atomic numbers. But if the hypothesis that an element of a certain atomic number exists is to be so formulated that it becomes testable, then much more is required than a 'strictly existential statement'. For example, the element with atomic number 72 (Hafnium) was not discovered merely on the basis of an isolated 'strictly existential statement'. On the contrary, all attempts to find it were in vain until Bohr succeeded in predicting several of its properties from his theory.

Yet Bohr's theory is a SYMO and *not* a 'strictly existential statement'. *Pace* Popper (1983, p. 70) and Olding (1984, p. 113) the discovery of data consistent with the SYMO for Hafnium did not confirm the 'existence' of Hafnium, i.e. the 'strictly existential statement', but confirmed claims of the validity of the SYMO for Hafnium. This confirmation in turn strengthened the (metaphysical) *belief* in the 'real existence' of an element of atomic number 72. In other words, confirmed SYMOs can lead to *beliefs* in the truth of certain 'strictly existential statements' about entities (whether these concern the existence of electrons of spin ½ or of Hafnium, cf. my earlier discussion). But *these beliefs do not amount to logical proofs or entailment of existence of these entities*, as I argued above against the logical argument of one referee of this paper, and as I have now argued against Popper and Olding.

Olding (1984, p. 109) apparently did not realize that 'strictly existential statements' are claims about the *existence* of systems, things or events (etc.). He believes that strictly existential statements 'are about particulars', but completely loses sight of the (metaphysically) claimed 'existence' of these particulars. This is not an accidental slip, but invalidates many of Olding's (1984) main arguments. For instance, Olding (1984, p. 115) claims (vis-à-vis Lewontin) that 'strictly existential statements' can 'be falsified by *true* universal propositions which imply their falsity' (italics are mine), a claim not made by Popper, who did not demand the *truth* of 'strictly existential statements-contradicting' universal propositions. Indeed, for reasons already stated, propositions of the type 'there exists *no X*' cannot be shown to be true, and 'strictly existential statements' are, accordingly, unfalsifiable. Moreover, as noted in relation to the example of Hafnium, it is SYMOs and not 'strictly existential statements' that can be confirmed or falsified by facts. Hence, 'strictly existential statements' are neither falsifiable nor confirmable, i.e. unlike SYMOs, they are *not testable*.

To compound his error concerning the alleged confirmability of a 'strictly existential statement' in the case of Hafnium, where *it was the* SYMO *which was confirmed* and *not* the 'strictly existential statement', Olding (1984, p. 111) makes a closely similar mistake. He speculates that I, G.D. Wassermann, would claim that a SYMO discussed by Schramm and Clayton (1978) 'is not a piece of science' because, according to Olding, the SYMO looks like a 'strictly existential statement', when it is nothing of the sort. So let us look at Olding's description of the theory concerned. He writes:

> There is a theory that a particular thing, the sun, was caused to exist by a particular event, a supernova (see Schramm and Clayton, 1978). According to this view a large star exploded about 4.6 billion years ago and the resulting shock-wave caused a nearby cloud of gas to contract to form the solar system. The theory was first suggested by the discovery of anomalous ratios of certain isotopes in a single meteorite which landed in Mexico, although these ratios have since been discovered in other meteorites of the general type known as carbonaceous chondrites... (Olding, 1984, p. 111).

Pace Olding, none of this statement is a 'strictly existential statement'. The sketched theory is a SYMO, and it refers to things that 'exist', and events that 'occur', *within* that SYMO, and refers to testable deductions from that SYMO. Existence of entities within a SYMO (e.g. point particles) does not amount to a claim of the existence of these entities as actual systems in the world. Here, again, we meet Olding's failure to distinguish SYMOs and 'strictly existential statements', an error also made by Popper (see above) and in a closely related form by one referee of an earlier version of this paper. (It is for this reason that I spent so much effort on the analysis of the present Section). Confirming Bohr's SYMO for Hafnium, or the SYMO discussed by Clayton and Schramm, by appropriate data consistent with these SYMOs, does not confirm 'strictly existential statements' about Hafnium or the supernova.

Typically, each SYMO depends (1) on a set of 'SYMO-specific *ad hoc* hypotheses' which particularize the SYMO, and (2) on a set of SYMO-specific variables which confer an element of *generalizing ability* on the SYMO. (Typical elementary examples of SYMOs in classical mechanics are the theory of the simple pendulum, the theory of a spinning top, and the theory of a massive point particle suspended by a vertical light elastic string which makes vertical oscillations (in a vacuum) about its position of equilibrium.) I shall show later that Olding (1984, pp. 110-111) has disregarded the generalizing element of SYMOs and has, instead, concentrated on particulars, wherever they occur in science, as if these particulars alone represented a balanced picture of approaches to systemic behaviour, which is not the case. In fact, the *deduction* of testable 'empirical generalizations$_{\text{deduced}}$' or of 'empirical laws$_{\text{deduced}}$' from SYMOs is only possible because of the presence of SYMO-specific variables. For instance, the solar *system*, or at least parts of it, can be modelled by appropriate SYMOs which, in good approximation, yield Kepler's laws as 'empirical laws$_{\text{deduced}}$'. Again, a metallic conductor of electricity is a system which can be modelled by an appropriate SYMO that yields Ohm's law as an 'empirical laws$_{\text{deduced}}$', and a large number of similar examples can be cited in the most varied branches of science. Thus, deduction of 'empirical laws$_{\text{deduced}}$' or 'empirical generalizations$_{\text{deduced}}$' from SYMOs forms an important aspect of scientific theorizing. In fact, one can say that a theory, e.g. a scientific hypothetico-deductive theory, '*explains*' an 'empirical generalization$_{\text{deduced}}$' or an 'empirical law$_{\text{deduced}}$' if an SYMO can be formulated within that theory which allows to deduce that 'empirical generalization$_{\text{deduced}}$' or 'empirical law$_{\text{deduced}}$', respectively. (As noted, near the start of Section 1, 'empirical laws$_{\text{deduced}}$' are special examples of 'empirical generalizations$_{\text{deduced}}$'). I shall discuss these topics further in detail in the following sections. (In Wassermann (1981, pp. 422-23) SYMOs were already introduced in all but name, and the need to deduce 'empirical generalizations$_{\text{deduced}}$' from SYMOs was stressed. Olding (1984, p. 109) totally misrepresented my earlier views (Wassermann 1974, 1981), as I shall show below, apart from failing to understand what scientific explanation is about.)

The second half of this paper deals, once more, like some of my recent papers (Wassermann 1978, 1981), with the role of theories in evolutionary science and with Olding's (1978, 1984) insistence that there could be evolutionary 'laws of succession'. I stated earlier (Wassermann, 1981, p. 429): "If, as Olding (1978) has claimed, there could be evolutionary laws rather than trends, then those who share his views ought to show what these laws are." Accordingly, Olding (1984) has tried to formulate several candidate 'evolutionary laws'. On inspection Olding's 'laws' turn out to be either quasi-tautologies or apparently untestable statements bordering on science fiction. Hence, there still exist no valid reasons for elevating evolutionary trends to the status of 'laws of succession'. Even the claim of Raup and Sepkoski (1984) to have detected a 26-million year periodicity in the mass extinctions of many organisms on Earth, remains not only controversial (cf. Hallam 1984; Maddox 1984), but does not really represent an evolutionary law, but a 'law of extinction'. I shall also stress, in general agreement with my earlier views (Wassermann 1981), that much of current evolutionary theorizing is concerned with evolution-specific SYMOs and the light which these cast on possible evolution-specific mechanisms.

2.3 Empirical generalizations, empirical laws and facts

Let me first turn to the role of facts in science. Facts are sometimes expressed by what Popper (1983) calls 'singular statements'. As is well known, scientific data are often obtained by means of complex observing machinery, which depends on one or more complicated theories of that machinery. Hence, indirectly, data themselves may be theory-related. Nevertheless, scientists often agree on what is to count and to be accepted as data or facts, even if there may be subsequent changes of opinion about the acceptability of previously accepted data. For instance, measuring techniques have vastly improved in many branches of science during the present century, and often called for revisions of, say, the accepted values of physical constants, or the precise structure of the DNA 'backbone' (etc.). Moreover, sometimes the search for specific kinds of data is stimulated by predictions of a scientific hypothetico-deductive theory (via a SYMO), while in other cases the invention of a new scientific hypothetico-deductive theory (and within it a particular SYMO) may be prompted by existing, but previously unexplained, data.

'Empirical laws', such as Ohm's law or Kepler's laws, are, as noted, special examples of 'empirical generalizations', and 'empirical generalizations' are *hypotheses* and universal statements. 'Empirical generalizations' can be obtained in one of two ways. (1) 'Empirical generalizations' can be *extrapolated* from data or facts. In this case I shall refer to them as 'empirical generalizations$_{\text{extrapol}}$' (and if the 'empirical generalization$_{\text{extrapol}}$' happens to be a 'law' it will be referred to as an 'empirical law$_{\mathrm{extrapol}}$'). 'Empirical generalizations$_{\mathrm{extrapol}}$' do not rely on isolated (single) data or events, but on *sets* of singular data, and extrapolation from sets of data to an 'empirical generalization$_{\mathrm{extrapol}}$' is invariably (at least implicitly *via* brain mechanisms) of a *statistical nature* (as explained in Wassermann (1982a)), e.g. based on curve-fitting in some cases (cf. Guest 1961; Wassermann, 1974, Chapter 2) or other statistical methods (cf. Wassermann 1982a), and these extrapolations go far beyond the data to which they are fitted. (In fact curves fitted to data need not contain any of the data points to which they are fitted, so that such fitted curves 'idealize' about the sets of data to which they are fitted.)

There are various philosophers (e.g. one referee of this paper) who believe that 'some genuine laws may only have *one* actual instance'. For example they believe that by spotting one black raven that this warrants the 'law' '*all* ravens are black'. I argued, however, elsewhere (see chapter 4) that such *non*-statistical extrapolations are the apparent roots of Hempel's paradox. Indeed, a sample of one instance does not warrant or permit a statistically formulated 'empirical law$_{\mathrm{extrapol}}$', and nonstatistical generalizations lead to absurdities, such as Hempel's paradox. I do not know of any accepted '*scientific* law$_{\mathrm{extrapol}}$' which is, or was, based on only one instance. *In fact, approximate repeatability, rather than isolated uniqueness of data is essential for the (statistical) attainment of useful* 'empirical laws$_{\mathrm{extrapol}}$'. In Section 2.4 of Wassermann (1974) I explained statistical extrapolations from data to 'empirical generalizations$_{\mathrm{extrapol}}$' in great detail (see also chapter 4) and placed much emphasis on the relation between data (i.e. certain kinds of 'particulars' in Olding's sense) and theoretical statements such as 'empirical generalizations$_{\mathrm{extrapol}}$'. Despite this, Olding (1984, p. 109), in one of his many wrong statements, claims that 'Wassermann's (1974) position is that pure science consists of theories about general classes of things although there are no logical objections raised against investigations of particulars'. Olding, apparently, has not read Section 2.4 of my 1974 book.

(2) As has been already repeatedly noted in this paper, often 'empirical generalizations' (or 'empirical laws') can be deduced from SYMOs of scientific hypothetico-deductive theories or other kinds of theories, and are then referred to as 'empirical generalizations$_{\mathrm{deduced}}$' (or 'empirical laws$_{\mathrm{deduced}}$', respectively). Sometimes discovery of an 'empirical generalization$_{\mathrm{extrapol}}$' precedes that of an 'empirical generalization$_{\mathrm{deduced}}$' ; sometimes it is the other way round. For instance, Kepler's laws were first put forward as 'empirical laws$_{\mathrm{extrapol}}$', and later Newtonian mechanics, combined with Newton's hypothesis of gravitation, allowed Kepler's laws to be deduced as 'empirical laws$_{\mathrm{deduced}}$'. In cases where an 'empirical generalization$_{\mathrm{deduced}}$' is established before an 'empirical generalization$_{\mathrm{extrapol}}$' of the related kind, the 'empirical generalization$_{\mathrm{deduced}}$' forms a *prediction* made by the SYMO from which it is deduced. Also, data are related

to 'empirical generalizations$_{extrapol}$' and sometimes to 'empirical generalizations$_{deduced}$'. One can only assess how well an 'empirical generalization$_{extrapol}$' fits an 'empirical generalization$_{deduced}$' or use an equivalent (*statistical*) procedure to test the 'goodness of fit' of an 'empirical generalization$_{deduced}$' and a set of data.

I shall now explain the relation between 'empirical generalizations' and SYMOs of scientific hypothetico-deductive (or other) theories, and base my account partly on a summary in my earlier paper (Wassermann, 1981, p. 422), although the present characterization goes beyond my earlier attempts (Wassermann 1974, 1981, 1982a) in important respects.

2.4 The link between scientific hypothetico-deductive theories and data

A typical hypothetico-deductive theory of science consists of the following ingredients among others.

(1) The scientific hypothetico-deductive theory postulates a set of *general starting hypotheses*' which apply to all members of the class of SYMOs and their mechanisms, with which that hypothetico-deductive theory deals. For instance, Newton's three laws of motion in classical mechanics (mainly the second and third laws), or Maxwell's equations in classical electromagnetic theory, are typical examples of 'general starting hypotheses' of scientific hypothetico-deductive theories. The 'general starting hypotheses' are either explicitly or implicitly formulated in terms of 'system-specific state descriptors' (for instance position coordinates and momenta of point particles in classical mechanics, state vectors in a Hilbert space in quantum mechanics, etc.). It is important to realize that 'general starting hypotheses' of scientific hypothetico-deductive theories (although occasionally called 'laws') are neither 'empirical generalizations' nor 'empirical laws' nor singular statements. 'General starting hypotheses' are, in fact, universal statements, but, unlike 'empirical generalizations$_{extrapol}$', they are not directly related to facts. 'General starting hypotheses' are only *indirectly* related to empirical states of affairs via 'empirical generalizations$_{deduced}$' that can be derived from the 'general starting hypotheses' when the 'general starting hypotheses' are *combined* with 'SYMO-specific *ad hoc* hypotheses' of a SYMO. A 'SYMO-specific *ad hoc* hypothesis' is *ad hoc* in the sense that it is specifically introduced for modelling some particular system. Thus, 'SYMO-specific *ad hoc* hypotheses' are *systematically particularized* (as was made quite clear in Wassermann (1981, p. 423) showing, *pace* Olding (1984), that even within scientific hypothetico-deductive theories one is concerned with general statements as well as with particularizing 'SYMO-specific *ad hoc* hypotheses').

(2) A scientific hypothetico-deductive theory invariably requires 'SYMO-specific *ad hoc* hypotheses' in addition to 'general starting hypotheses' in order to specify any particular SYMO. 'SYMO-specific *ad hoc* hypotheses' invariably idealize (as do 'general starting hypotheses') and it is this which makes SYMOs assumed idealized models of actual systems (i.e. systems that people *believe* to 'exist' in 'reality'). For instance, when one assumes that the string of a 'simple pendulum' (the theory of the 'simple pendulum' being a typical SYMO) is 'inextensible' and 'light', that the bob of the pendulum can be treated as a 'point particle', that air resistance can be neglected, that the system makes 'small' oscillations in a vertical plane about its position of stable equilibrium (relative to the earth's surface), that specified initial conditions prevail, then one is only invoking a set of 'SYMO-specific *ad hoc* hypotheses' (in this case pendulum-specific idealizations) for the SYMO concerned. Similar remarks apply to innumerable other SYMOs of the same or other scientific hypothetico-deductive theories.

(3) One must try to deduce as many 'empirical generalizations$_{\text{deduced}}$' (or 'empirical laws$_{\text{deduced}}$') as possible from the combination of 'general starting hypotheses, and SYMO-specific *ad hoc* hypotheses' of the SYMO used. (For instance, in the case of the simple pendulum one has deduced the period of oscillation as a function of the length of the string and other constants as an 'empirical generalization$_{\text{deduced}}$'). When making these deductions it is often expedient to rely on mathematical approximation procedures, where rigorous deductions are not possible.

(4) One asserts that a scientific hypothetico-deductive theory, via a SYMO, '*explains*' facts or data either if an 'empirical generalization$_{\text{deduced}}$', deduced from that SYMO, 'fits' data in accordance with a *publicly agreed* criterion of 'goodness of fit', or if an 'empirical generalization$_{\text{deduced}}$', deduced from that SYMO, fits sufficiently closely (according to an *accepted criterion*) some existing 'empirical generalization$_{\text{extrapol}}$'. Conversely, *every set of facts that is 'explained' via an* SYMO, *in the manner just stated, serves to confirm (or strengthen) that* SYMO. If an 'empirical generalization$_{\text{deduced}}$' *predicts* an 'empirical generalization$_{\text{extrapol}}$' (i.e. if it precedes the discovery of data that lead to that 'empirical generalization$_{\text{extrapol}}$'), then the prediction may sometimes be taken as evidence for generating belief in the 'existence' of a previously unknown 'system'. For instance, Dirac's SYMO of the 'spinning electron' led to 'empirical generalizations$_{\text{deduced}}$' that were interpreted as evidence for belief in the 'existence' of entities corresponding to the hypothetical construct called the 'positron'. Similarly, when the theory of electro-weak interactions (a scientific hypothetico-deductive theory, cf. Taylor, 1976) predicted, via a SYMO, particular 'empirical generalizations$_{\text{deduced}}$', these were interpreted as properties of 'hypothetical

constructs' (or 'systems') that are believed to 'exist' and were then called *W*- and *Z*-bosons. Subsequently appropriate 'empirical generalizations$_{extrapol}$' *confirmed* the validity or acceptability of the theory-derived 'empirical generalizations$_{deduced}$' that related to a belief in these bosons, thereby *strengthening the belief in the 'existence' of these bosons,* which were then claimed to have been 'discovered'.

I conclude that the fitting of data by an 'empirical generalization$_{deduced}$' (either directly or via an 'empirical generalization$_{extrapol}$') is *the only way in which a scientific theory (say a scientific hypothetico-deductive theory) can, via a* SYMO, *'explain' data and also the only way in which data can confirm a* SYMO. If Olding (1984, p. 112) complains that Popper's methodology does not guarantee the 'truth' of theories (or even their 'probability', see also Wassermann (1985)), then this is so because Olding has not grasped in which sense a scientific theory 'explains' facts. As stated in Section 2, SYMOs do not pretend to be 'revealed truths', but are makeshift models, and the failure to demonstrate their 'truth' (because they are not 'true', unless by accident unknown to us!) does not make all scientific theories untestable (see Olding's 1984, p. 112 claims to the contrary, notably line 6 and elsewhere on that page).

(5) In his tirade against Popper's methodology (and indirectly against the methodology of most or all practicing scientists) Olding (1984, p. 112) remarked that

> It follows immediately that if we have no idea whether or not the theories we use to 'explain' particular events...are true or probable, then the only possible sense of 'explain' is a conventionalist one. There can be no hope of explaining events in the realist sense of, for example, showing why they are as they are, or even that they have occurred at all and the only thing left to do, unless one gives up science, is to arrange one's universal and particular propositions in a certain kind of order. It is no wonder that, in their dispute with evolutionists, 'creationists' have found a friend in Popper: they simply arrange their biological propositions in a different order but with a little help from God and the Bible.

Pace Olding, scientists have for generations produced scientific hypothetico-deductive theories and SYMOs and derived from the latter 'empirical generalizations$_{deduced}$' which they compared with facts or with 'empirical generalizations$_{extrapol}$' without worrying about the absolute 'truth' of scientific hypothetico-deductive theories, and SYMOs were never meant to be 'true'. Just the same, science has flourished.

(6) SYMOs need neither be true, nor do they represent 'ultimate realities' of systems or guarantee their 'existence', as was explained in Section 2.2. In this

context I wrote (Wassermann 1974, p. 106, now somewhat altered to suit the present nomenclature and argument):

> The modelling of a 'hypothetical construct' can change if, as time progresses, the construct retains its name but its model becomes based on a different SYMO of the same or a different hypothetico-deductive theory. For example, in J.J. Thomson's time the hypothetical construct 'electron' was modelled in terms of Maxwell's and Newtonian theories...Later Bohr's SYMOs remodelled the concept 'electron' by using the original SYMOs (in the Hamilton-Jacobi version) jointly with supplementary quantization rules (cf. Wilson 1940, Vol. 3). Subsequently Schrödinger's equation, together with the probabilistic interpretations of non-relativistic quantum mechanics, provided yet other remodellings (of the hypothetical construct 'electron' via yet different SYMOs). Then came Dirac's (cf. 1958) relativistic equation of the electron, another SYMO (together with its deductions and interpretations). And this process may go on, so that the hypothetical construct 'electron' may mean something different to people using different theories (i.e. different SYMOs of different hypothetico-deductive theories).[1]

These successive uses of different hypothetico-deductive theories and SYMOs, for explaining data associated with the same 'hypothetical construct', are not anarchical, but were guided by many considerations of plausibility (e.g. Bohr's correspondence principle in quantum mechanics, etc.). Yet, they serve to show that people are sometimes driven to invent new hypothetico-deductive theories with new SYMOs, so as to explain either new data or to explain known data better than previous SYMOs of preceding hypothetico-deductive theories. This, however, does not mean that a given hypothetico-deductive theory may not survive indefinitely, since many of its SYMOs may remain applicable, (e.g. the SYMOs of classical mechanics remain applicable to many systems of pure and engineering science, etc) and new SYMOs of that theory may still be formulated. There exist also many examples of theorists having used different (usually less idealized and more 'realistic') SYMOs to rework the theory of the 'same' type of system within the same hypothetico-deductive theory. None of these SYMOs are 'true' or correspond to 'ultimate realities' (unless one of these SYMOs *by chance*, and unknowable to us, *is* 'absolutely' true and represents an 'exact' model of a 'real' system).

I conclude, incidentally, from the preceding arguments, that Olding's view of what science should be like does not correspond to scientific practice or to the recorded history of science. His notions of the nature of scientific 'explanation' as having to be truth- and reality-bound simply cannot be satisfied, and are not required by SYMOs and the scientific methodology based on SYMOs and their ways of 'explaining' facts.

(7) Many theorists establish 'Mini-Theories', entirely based on 'SYMO-specific *ad hoc* hypotheses'. Such theories, although they are SYMOs, do not rely on hypothetico-deductive theories and their 'general starting hypotheses'. I gave an example of this elsewhere (Wassermann 1981) and numerous examples of typical 'mini-theories' can be found in the *Journal of Theoretical Biology* and in the *Bulletin of Mathematical Biology*.

(8) Many systems considered by scientists are members of classes of systems, and the SYMOs formulated are then assumed to be members of the class of SYMOs for the class of systems concerned, subject, of course, to use of 'SYMO-specific *ad hoc* hypotheses' for each SYMO concerned. Applied scientists also formulate SYMOs for unique systems (with the help of appropriate 'SYMO-specific *ad hoc* hypotheses', and generalizing variables). *In general then science is primarily concerned with facts,* SYMOs *and regularities expressed as 'empirical generalizations' (or 'empirical laws') where possible.* Often systems appear in repeat editions (e.g. hydrogen atoms) in the universe or on Earth. The principal keys to understanding and 'explaining' the behaviour of systems are always SYMOs and their related 'empirical generalizations$_{\text{deduced}}$'. *Pace* Olding (1984, p. 110) nobody disputes that in some sense the solar system is a particular or a 'unique' system. This, however, has not debarred us from establishing SYMOs for parts of that system, and allowed to deduce Kepler's laws (as 'empirical laws$_{\text{deduced}}$') for the motion of the planets, and to explain many other properties via such SYMOs. Again, although the whole universe is a unique system (as far as we know!), this has not stopped us from formulating cosmological (etc.) models (SYMOs) in attempts to explain regularities of that system (e.g. its expansion [Hubble's law], the nature of its subcomponents, galaxies, stars, etc). What matters to scientists is not whether systems are unique or appear in multiple copies, but whether people can *explain* (and *predict*, see below) the modes of workings of systems via SYMOs in sufficient approximation, i.e. whether they can explain their *mechanisms* and whether they can produce testable and predictive 'empirical generalizations' for systems.

(9) People treasure hypothetico-deductive theories and other scientific theories (e.g. 'mini-theories') not just because they 'explain' facts in the sense mentioned, but also because SYMOs yield *predictions* (e.g. previously unknown 'empirical generalizations$_{\text{deduced}}$'), and likewise, 'empirical generalizations$_{\text{extrapol}}$' allow us to predict. In fact, in some branches of science one has yet no hypothetico-deductive theories or even mini-theories, but has only advanced to the stage of formulating 'empirical generalizations$_{\text{extrapol}}$'. (Strangely enough, Olding (1984) did not discuss the predictive aspects of science, possibly because

of his unjustified preoccupation with the alleged need for 'truth' and 'reality' status of theories; see above.)

2.5 Evolution-specific SYMOs, evolution-specific mechanisms and laws in evolutionary science

I shall now apply some of the conclusions of the preceding sections to theoretical evolutionary sciences. Many evolutionists, in perfectly respectable journals and books, have indulged in pure speculations by providing 'explanations' of evolutionary adaptations based on ad hoc constructed 'strictly existential statements' (not to be confused with SYMO-specific *ad hoc* hypotheses' of SYMOs, a confusion that surfaces in Olding's (1984, p. 111) paper). Typical speculations of this kind may be of the form (presumably fictitiously invented) exemplified by Lewontin (1972) that "a species of bird with a small bill may evolve a large bill size, because some aspect of the environment has changed so that large-billed birds now have more offspring". Indeed, I have seen many similar examples in the literature of the subject. The example cited implicitly postulated the existence of a (selective) environmental factor, and then asserts, *without supplying a* SYMO *of how this factor could have operated*, that 'because' of the *existence* of this factor a particular adaptation arose (see also Wassermann, 1978). Lewontin (1972) asked pertinently:

Can one really imagine observations about nature that would disprove natural selection as a cause of different bill size? The theory of natural selection [as abused in this context, where it depends on a hidden ad hoc 'strictly existential statement' in the sense of Section 2.2 above] is then revealed as metaphysical, rather than scientific. Natural selection [when invoked in this fashion via an ad hoc 'strictly existential statement'] explains nothing because it explains everything.

Among distinguished evolutionists Lewontin is not alone in reaching this conclusion. The equally distinguished evolutionist N. Eldredge, in a book review (1982, *Nature Lond.* **296**, p. 508) of the philosopher Michael Ruse's (1982) book *Darwinism Defended* (Reading, Mass.: Addison-Wesley), noted that

> Nowhere in this single-minded tract is Ruse's dedication to the almighty principle of natural selection better displayed than in his little section on tribolite vision. The mathematically 'perfect' shape of phacopid and dalmanitid lenses is, to Ruse, exquisite evidence of the power of natural selection. I confess I cannot fathom the difference between Ruse's argument and the older creationist argument from design: see this organ system; observe its intricacy! Only (God, natural selection) could have fashioned

> such a marvellous organic machine! There is a difference, of course: God, as a supernatural being, does not belong to science, whereas natural selection patently does. But used in this inappropriate fashion, natural selection becomes a mere substitute for the Creator. It tells us nothing, really, about tribolite eyes or anything specific or meaningful about how they came into existence.

Such panselectionist abuses of the notion of 'natural selection' seem also hinted at in Waddington's (1967, p. 97) remarks that

> Some physicists were saying that [the theory of evolution by natural selection] is not a real theory at all, either it does not explain anything or it explains everything ... The criticism was that if any animal evolves one way biologists have a perfectly good explanation [based on unfalsifiable ad hoc 'strictly existential statements' referring to unknown 'selective factors']; but if it evolves some other way, they have an equally good explanation [based on different unfalsifiable ad hoc 'strictly existential statements' referring to different unknown 'selective factors']. So what is the good of all this explanation?

Olding (1984, p. 109), while quoting me out of context, was careful not to mention the preceding conclusions by Waddington and Lewontin (quoted explicitly in my 1978 paper) and accused me, instead, of rejecting speculations based on ad hoc 'strictly existential statements' (such as 'there exists *a* selective factor' (which remains unspecified and unknown)), which are published in 'reputable science journals' as 'merely metaphysical', as if this were not in line with what Lewontin and Waddington (and more recently Eldredge) complained of.

While, unlike Olding (1984), many evolutionary theorists have realized that evolutionary explanations cannot be furnished by ad hoc 'strictly existential statements', even if they appear in 'reputable journals' many of these theorists (e.g., Mayr 1963; Lewontin 1974; Maynard Smith and Savage, 1956) have developed SYMOs, namely *evolution-specific* SYMOs, in order to discuss evolution-specific mechanisms. For instance, Mayr developed important evolution-specific SYMOs for speciation, the 'system' modelled being a species. Thus by deed, perhaps more than by explicit philosophy, leading evolutionary theorists have realized that what counts in theoretical science are SYMOs and their deductions, and *not* ad hoc 'strictly existential statements'. There exist now various evolution-specific SYMOs for speciation (Mayr 1963; White 1978), evolution-specific SYMOs for population genetics (Lewontin 1974; Kimura 1979a, 1979b), evolution-specific SYMOs for adaptation (Wassermann 1982b, 1982c, 1997) and so forth. Thus, I stated (Wassermann, 1981, p. 419) that the

'Theory of Evolution' is to be understood as a *hypertheory* which comprises (1) the set of all evolution-specific SYMOs that provide evolution-specific mechanisms, (2) the set of hypotheses that postulate how members of various classes of viable evolution-specific mechanisms could have been interrelated within the course of evolution and (3) deductions from the evolution-specific SYMOs and interrelating hypotheses (I have slightly changed the nomenclature of Wassermann (1981) to correspond to that adopted in the present paper).

The trouble with evolution-specific SYMOs is that they do not easily, if at all, in their present states of development, enable us to derive any testable 'empirical laws' (or 'empirical generalizations') in the sense in which SYMOs of physics or other 'exact' sciences permit us to derive data-fitting 'laws'. The reason is that current evolution-specific SYMOs use *highly idealized* 'SYMO-specific ad *hoc* hypotheses'. Accordingly, one can only test some of the evolution-specific SYMOs *qualitatively* (cf. Wassermann 1978, pp. 227 ff). This, at least, is the present situation. In fact, whereas some proponents of evolution-specific SYMOs assume that selection, by cumulative assembly of suitable alleles in populations, 'created' all new genotypes and corresponding phenotypes (the 'panselectionist view'), others, while not denying the existence of a measure of selection (much of which is in any case 'purifying', i.e., 'eliminative') do not think that selection is as prominent as panselectionists believed (cf. Kimura's 1979a, 1979b SYMOs), and many adaptations could have been produced by other processes in addition to mutations and selection (cf. Wassermann 1982b, 1982c, 1997 for suggested SYMOs for molecular adaptations). At any rate, it is hard enough to test a few of the existing evolution-specific SYMOs of population genetics qualitatively (cf. Wassermann 1978), let alone test how such evolution-specific SYMOs operated throughout the course of evolution in creating organs and other adaptations of organisms.

The high idealization of 'SYMO-specific *ad hoc* hypotheses' of evolution-specific SYMOs has, I believe, something to do with the *heterogeneity* of composition of the systems dealt with by evolutionists, whether these systems are chromosomes, cells, individual organisms, or populations. 'Empirical laws', like those of Boyle, Ohm or Kepler, are only possible because one is either (1) dealing with (for some purposes) *simple* systems (like the solar system) that behaved as if they were quasi-isolated and showed repetitive aspects, at least when time-averaged during human historical time spans, or is (2) dealing with systems made up of many copies of repeat units, such as the molecules of a simple gas (in the case of Boyle's law) or an electric conductor containing approximately repeat crystalline units of metallic ions and interacting electrons (in the case of Ohm's law). In the case of repeat units one can often use *statistical mechanics*. Even the basic structure of the DNA backbone (but not the side chains) shows a repeat pattern. By contrast, one cannot apply anything corresponding to statistical mechanics (*pace* some applications of the Fokker-Planck equation in highly idealized population genetic models) to assemblies of strongly heterogeneous systems, unless one over-idealizes and regards systems as being alike that really differ in significant ways (e.g., Olding's implicitly assumed many similar generations of apple trees; see below). It is for this reason that although one can establish SYMOs for a unique system, like the *'whole universe'*, and deduce Hubble's law, because this whole system consists, in the idealization used, of similar subunits (galaxies), one cannot aspire to discover a SYMO for the totality of living organisms on Earth (including all eco-systems) and derive 'laws of succession' from such a SYMO. The best that one can hope to discover for such heterogeneous systems are certain long-term trends for certain classes of such systems (e.g., species, populations, etc.) and obtain these trends empirically but not as deductions from SYMOs.

It is for these reasons that I consider it futile to look for 'laws of succession' in evolution (as distinct from trends). Instead, as working evolutionists have recognized, it is better to establish greatly simplified SYMOs and test, as far as possible, on a qualitative basis, some restricted aspects of these SYMOS. Since Olding (1978, 1984) thinks otherwise, let us look at some of his alleged 'laws of succession'. Olding (1978, p. 137) wrote: 'Given a chance apple trees regularly produce apples and not plums or brussel sprouts. There are here regular sequences which, allowing for the usual vagaries of nature, can be caught in laws of succession'. This point is repeated in Olding's (1984, p. 109) paper. He does not seem to realize that if organisms of a particular genotype always produced the same *genotype* then there could be no speciation. The underlying heterogeneity is obvious: phenotypically similar appearing apple trees of the same species may differ in DNA composition. That apple trees produce apples and not plums is a trend which could change for *some* apple trees by the long-term operation of evolution-specific mechanisms of speciation, these changes representing unpredictable trend-altering contingencies. I am not referring here to major destructive contingencies such as mass exterminations of organisms, but to relatively slow 'creative' contingencies. The possibility that evolution-specific mechanisms of speciation may be at work, unknown to us, is always present (even in apple trees). New evolutionary diversity crops up all the time, as unpredictably as long-term weather trends. It creates unpredictable diversifications of gene pools, so that the fruit of apple trees, although not looking like brussels sprouts, need genetically not precisely resemble the fruit of immediate ancestors.

Thus, far from having established a 'law of succession', Olding has produced a near-tautology. His phrase (above) 'given a chance' amounts to saying that in the absence of species-generating (genotype-changing) evolution-specific mechanisms there will be no change of genotype leading to speciation. In fact, one major reason why contingencies (other than catastrophes) have been important for evolution, and rule out 'laws of succession' (at least at the genetic level), is the existence of millions of species. Surprisingly, Olding (1984, p. 106) has realized 'that the way specific lineages of organisms have evolved depends on too many contingencies to be caught in laws, as Wassermann and Popper have argued (cf. Olding 1978, p. 141) ...'. Yet Olding failed to appreciate that the contingencies present in speciation also invalidate his apple tree 'law of succession' as an 'empirical law' rather than a trend. What seems to appear *repeatedly* in evolution is the operation of *similar* SYMOs (and evolution-specific mechanisms), e.g. SYMOs of speciation, which may generate trends (as noted in Wassermann, 1981).

In addition to the wrong 'law of succession' for apple trees, Olding (1984) also proposed a 'law of succession' based on Whittaker's (1969) classificatory scheme for organisms. The details of Whittaker's classifications are irrelevant for the following discussion, since Olding seems to be in trouble on quite different grounds. He wrote (Olding, 1984, p. 105-106):

> We might find exemplification of evolutionary laws, genuine laws of succession, in Whittaker's scheme. One, perhaps fundamental, law might be wherever large-scale evolution occurs (and we might add appropriate qualifications about the kind of planet supporting it, and so on) it will always go through three stages corresponding to Whittaker's level of organization and will branch out into similar nutritive ways of life.

Since Olding's new 'law of succession' explicitly envisages that what could happen on some *other* planets could be similar (as regards *major* stages) to what *has already happened on Earth*, Olding's 'law' could only be *tested* provided there are other similar planets in our galaxy on which life flourishes and with which we could communicate. Indeed, not long ago it was widely held that life on Earth is only one instance of life on myriads of satellites of appropriately sized stars in our galaxy or other galaxies. This belief is no longer held by many, or most, scientists (cf. Rood and Trefil (1981) and O'Neill's (1981) summary of their conclusions). O'Neill (1981) stressed that there is a strong possibility that life is unique to our planet. He referred to Michael Hart, who

> set up a mathematical model of the Earth atmosphere, and in 1978 traced the evolution of that atmosphere by computer over geological time from its formation by volcano emission to its complete transformation through biochemical processes once life had formed. Hart found that the survival and evolution of life depended on an extra-ordinarily lucky series of accidents, by which the temperature of the Earth remained remarkably constant for

most of the age of our planet. Had we not been so fortunate as to be located at precisely the right distance from the Sun, within margins of no more than two per cent, Earth would now be as lifeless as Mars or Venus, according to Hart's calculations. (See also Trefil, 1982).

If Hart's model is valid, then Olding's predictions concerning the applicability of Whittaker's three stages to *other* planets, endowed with life, would be untestable if there should be no other planets in our galaxy endowed with life. Still, Olding could argue that Hart's calculations are based on a fallible SYMO which could be unreliable.

Another argument, used by O'Neill (1981) and others (starting with Fermi many years ago), is based on the belief that if life existed anywhere and led to civilizations more advanced than ours on other planets, then members of these civilizations would have sent self-replicating robot probes to other planets like ours. That no such robot probes (or even more daringly their owners) have been discovered on Earth speaks against the existence of such civilizations, so it is argued. There remains, however, the possibility that if other civilizations like ours have existed before us in our galaxy, then, before being able to send out self-replicating probes, they committed thermo-nuclear suicide, and wiped out all life on their planets while doing so. If this is a 'law of evolution', then it would provide good reasons why *we* shall never be able to test Olding's law. In any case, such considerations, however speculative, belong less to the realm of science fiction than Olding's extrapolation of Whittaker's findings to other planets as a 'law of succession', and suggest that Olding's extrapolations are likely to be *untestable*, so that his second proposed 'law of succession' is likely to be metaphysical (in Popper's sense).

Olding (1984) has repeatedly drawn attention to the increasing complexity of organisms during evolution (see his pp. 106-108). This *trend*, however, could be due to the operation of evolution-specific mechanisms for adaptation (cf. Wassermann 1982b, 1982c, 1997 for candidate evolution-specific mechanisms), and the possibility of the discovery of such further evolution-specific mechanisms is even conceded by Olding (1984, np. 107). The second law of thermodynamics, which states that for a closed (isolated) system entropy can never decrease, is not too relevant, since it does not rule out *localized* long-term decreases of entropy *within* that system. There exist also entropy increasing trends in our solar system that will ultimately terminate all life on Earth, if it is not terminated by ourselves first. Thus, evolutionary trends remain trends, and must not be confused with indefinitely operating 'laws of succession', which do not seem to exist.

2.6 Conclusion

It is argued that theoretical scientists try to establish *scientific systemic models* (SYMOs) based on (1) 'General starting hypotheses' in the case of scientific hypothetico-deductive theories, and on (2) 'SYMO-specific *ad hoc* hypotheses' which specify a SYMO in very idealized ways, and which include initial conditions and other SYMO-specific ad hoc assumptions. (A typical elementary example of a SYMO is the theory of the 'simple pendulum' in mechanics.) Partly because of their idealizations, SYMOs do not claim to represent 'reality' of systems or to be 'true' theories. It is argued at length that the nature of the ultimate 'reality' of a system, and the actual mode of 'existence' of that system, is unknowable (except that by chance, and unknown to us, a SYMO could reveal an absolute truth about a system). SYMOs and other brain-representation-linked *models* serve only as means of *representing* supposedly real systems. Yet, models must not be confused with the systems which they model. Many SYMOs have yielded 'empirical generalizations$_{\text{deduced}}$' or 'empirical laws$_{\text{deduced}}$' which were *deduced* from the SYMOs and can be compared with data either directly or via 'empirical generalizations$_{\text{extrapol}}$' or, respectively, via 'empirical laws$_{\text{extrapol}}$' which are *extrapolated* from sets of data. This explains the nature of scientific 'explanations' via SYMOs. Evolutionists have developed various kinds of evolution-specific SYMOs for discussing evolution-specific mechanisms, but these SYMOs have yielded only weakly *qualitatively* testable 'empirical generalizations$_{\text{deduced}}$', and reasons for this are given. Olding's claim to have discovered 'evolutionary laws of succession' is examined. Olding's 'laws' are either quasi-tautologies or apparent metaphysical (untestable) assertions. Olding's (1984) many methodological arguments against Popper's, Lewontin's, and the author's positions are examined and are found to represent a sizeable number of serious philosophical errors.

Note

1 In fact, quite recently 'electrons' and other kinds of elementary 'entities', previously interpreted (within SYMOs) as 'point particles' have become reinterpreted as 'superstrings' (cf. M.B. Green (1985), *Nature Lond.* **314**, p.409-414; J. Ellis (1986), *Nature Lond.* **323**, 595).

References

Dirac, P.A.M.: 1958, *The Principles of Quantum Mechanics* (4th edn.), Clarendon Press, Oxford.

Guest, P.G.: 1961, *Numerical Methods of Curve Fitting*, Cambridge University Press, Cambridge.

Hallam, A.: 1984, 'The Causes of Mass Extinctions: *Nature Lond.* **308**, 686-87.

Kimura, M.: 1979a, 'Model of Effectively Neutral Mutations in which Selective Constraint is Incorporated', *Proceedings of the National Academy of Science USA* **76**, 3440-44.

Kimura, M.: 1979b, 'The Neutral Theory of Molecular Evolution', *Scientific American* **241**, 94-104.

Lewontin, R.C.: 1972, 'Testing the Theory of Natural Selection', *Nature Lond.* **236**, 181-82.

Lewontin, R.C.: 1974, *The Genetic Basis of Evolutionary Change*, Columbia University Press, New York.

Maddox, J.: 1984, 'Extinctions by Catastrophe?' *Nature Lond.* **308**, 685.

Maynard Smith, J. and R.J.G. Savage: 1956, 'Some Locomotory Adaptations in Mammals', *Zoological Journal Linnean Society* **42**, 603--22.

Mayr, E.: 1963, *Animal Species and Evolution*, The Belknap Press of Harvard University, Cambridge, Mass.

Oddie, G.: 1981, 'Versimilitude Reviewed', *British Journal for the Philosophy of Science* **32**, 237-65.

Olding, A.: 1978, 'A Defense of Evolutionary Laws', *British Journal for the Philosophy of Science* **29**, 131-43.

Olding, A.: 1984, 'On Theories of Evolution and Their Several Natures', *Nature and System* **6**, 103-17.

O'Neill, G.K.: 1981, 'Where is Everybody? Some New Answers', *Nature Lond.* **294**, 25.

Popper, K.R.: 1983, *The Logic of Scientific Discovery* (11th impression), Hutchinson, London.

Raup, D.M. and J.J. Sepkoski: 1984, 'Periodicity of Extinctions in the Geological Past', *Proceedings of the National Academy of Science USA* **81**, 801-05.

Rood, R.T. and J.S. Trefil: 1981, *Are We Alone? The Possibility of Extraterrestrial Civilizations*, Charles Scribner's Sons, New York.

Schramm, D.N. and R.N. Clayton: 1978, 'Did a Supernova Trigger the Formation of the Solar System?', *Scientific American* **239**, 98-113.

Taylor, J.C.: 1976, *Gauge Theories of Weak Interactions*, Cambridge University Press, Cambridge.

Trefil, J.S.: 1982, 'Where are They now?', *Nature Lond.* **18**, 128-39.

Waddington, C.H.: 1967, 'General Discussion' (p. 97), in P.S. Moorhead and M.M. Kaplan (eds.), *Mathematical Challenges to the Neo-Darwinian Interpretation of Evolution*, The Wistar Institute Press, Philadelphia.
Wassermann, G.D.: 1974, *Brains and Reasoning*, Macmillan, London
Wassermann, G.D.: 1978, 'Testability of the Role of Natural Selection Within Theories of Population Genetics and Evolution', *British Journal for the Philosophy of Science* **29**, 223-44.
Wassermann, G.D.: 1981, 'On the Nature of the Theory of Evolution', *Philosophy of Science* **48**, 416-37.
Wassermann, G.D.: 1982a, 'On the Laws of Nature', *Synthese* **51**, 381-96.
Wassermann, G.D.: 1982b, 'TIMA Part 1. TIMA as a Paradigm for the Evolution of Molecular Complementarities and Macromolecules', *Journal of Theoretical Biology* **96**, 77-86.
Wassermann, G.D.: 1982c, 'TIMA Part 2. TIMA-based instructive Evolution of Macromolecules and Organs and Structures', *Journal of Theoretical Biology* **99**, 609-28.
Wassermann, G.D.: 1985, 'On the Nature of Inductive Probabilities', *Methodology and Science*, **18**, 128-39.
Wassermann, G.D.: 1997, 'TIMA Part 3', In *Keys to Life*, Ashgate Publishing Ltd, Aldershot.
White, M.J.D.: 1978, *Modes of Speciation*, W.E. Freeman & Co., San Francisco.
Whittaker, R.H.: 1969, 'New Concepts of Kingdoms of Organisms', *Science* **163**, 150-60.
Wilson, W.: 1940, *Theoretical Physics*, Methuen, London.

The University of Newcastle upon Tyne
Newcastle upon Tyne
NE1 7RU
United Kingdom

3 Occam's razor: a blunt instrument

There are many scientists, philosophers and laymen, who, misguidedly, believe that the so-called 'Occam's Razor', also known as 'The Principle of Parsimony of Hypotheses' forms an integral part of scientific and other kinds of theorizing. I propose to show that Occam's Razor, which, at best, is an arbitrary convention, turns out, on closer examination, to be an absurdity.

It is widely believed that Occam's Razor applies particularly to the most 'majestic' scientific hypothetico-deductive theories, such as Classical Mechanics, Classical Electromagnetic Theory, Theoretical Optics, Equilibrium (and non-equilibrium) Thermodynamics, Quantum Mechanics, Quantum Electrodynamics and Quantum Chromodynamics, among others. (The situation as regards the Theory of Evolution is somewhat different[1].) I have recently provided a new formalization of the characteristic structure of most, or all, typical scientific hypothetico-deductive theories in a paper[2] (and see the preceding chapter) and a book[3]. In order to show the cardinal weakness of Occam's Razor let me recapitulate briefly the essentials of my formalization of scientific hypothetico-deductive theories. One could hardly expect that William of Occam could have known or anticipated any of this, since he lived in the first half of the 14th century. What is worse, is that many twentieth century scientists and philosophers have adopted, or try to use still blindly, Occam's Razor, without realizing its demonstrable failure in the case of many of the most prominent scientific hypothetico-deductive theories.

A typical scientific hypothetico-deductive theory has a number of starting hypotheses (not to be confused with initial conditions of SYMOs). For instance, Newton's three laws of motion are starting hypotheses of Newtonian Mechanics and Maxwell's equations are starting hypotheses of Classical Electromagnetic theory. The starting hypotheses are then applied to a great variety of different kinds of systems. For each system a systemic model (SYMO) is constructed (if possible), by combining the starting hypotheses with SYMO-specific *ad hoc* hypotheses.[4] A typical SYMO, known to many school children, is the theory of the simple pendulum in Classical Mechanics, and its SYMO-specific *ad hoc* hypotheses are, among others, the assumptions that the bob of the pendulum is a 'point particle' and that the string of the pendulum has negligible (i.e. zero) mass. When the SYMO-specific *ad hoc* hypotheses and the starting hypotheses of the scientific hypothetico-deductive theory are combined, then it is often possible (e.g. with the help of mathematics) to make testable deductions.

Within a scientific hypothetico-deductive theory, such as Newtonian Classical Mechanics[5] there exist literally hundreds or, perhaps, thousands of different kinds of SYMOs (many of which can be found deposited in patent offices or technical libraries and quite a good many in undergraduate texts). Since each SYMO involves one, or, usually, several SYMO-specific *ad hoc* hypotheses, Classical Mechanics comprises hundreds or, perhaps, *thousands* of different kinds of SYMO-specific *ad hoc* hypotheses. Similar remarks apply to the other scientific hypothetico-deductive theories mentioned.

Hence, if a prominent theory depends on hundreds or thousands of *ad hoc* hypotheses, then this makes a nonsense of Occam's Razor in the twentieth century. In fact, theories such as Newtonian Mechanics are open-ended, and can be applied (and are applied) to ever new systems, so that ever new *ad hoc* hypotheses are added to these theories. What is true, however, is that in some cases one may have two theories with competing hypotheses which try to explain the same set of phenomena. In such cases Occam's Razor stipulates that the theory with the fewest hypotheses is adopted. A typical example is the pre-1926 quantum theory of Bohr, Wilson and Sommerfeld which, perhaps, had more hypotheses than modern quantum mechanics. Yet, the earlier theory was ousted, not because it had more hypotheses, but because it gave many wrong results. Thus, it is usually the validity of a theory rather than the number of its hypotheses which determines its survival. Some or even many twentieth century scientists and philosophers do not seem to have appreciated the state of affairs just described, because they have been indoctrinated about the spurious 'value' of Occam's Razor.

To sum up. If a prominent and far-reaching scientific hypothetico-deductive theory, such as classical mechanics, invokes hundreds, or, more likely, thousands of SYMO-specific *ad hoc* hypotheses (see chapter 2 above), using appropriate ones of these hypotheses for each of hundreds or thousands of different SYMOs (= systemic models), then it is preposterous to claim that such a theory is parsimonious in hypotheses, unless the word 'parsimony' loses its usual meaning.

Notes

1 See G.D. Wassermann, 'On the Nature of the Theory of Evolution' *Philosophy of Science 48* (1981) 416-437

2 G.D. Wassermann, 'Theories, Systemic Models (SYMOs), Laws and Facts in the Sciences' *Synthese 79* (1989) 489-514, or chapter 2 of this book.

3 G.D. Wassermann, *A Philosophy of Matter and Mind* (Aldershot: Ashgate Publishing, 1994)

4 Initial conditions and boundary conditions of SYMOs are examples of SYMO-specific *ad hoc* hypotheses.

5 Corresponding remarks apply to the Hamiltonian and Lagrangian versions of classical mechanics.

4 On the laws of nature

4.1 Introduction

All so-called 'laws of nature' are scientific hypotheses. There exist, however, two drastically distinct major classes of scientific hypotheses some of whose members have been referred to as 'laws of nature'. It is the differences and relations between these two classes of hypotheses that I shall try to examine in this paper. Possibly because the word 'law' may convey some kind of necessity, contemporary scientists, who have extensively realized the fallibility or tenuous nature or severe limitations of all of their hypotheses, have shrunk back from calling more recently propounded scientific hypotheses 'laws of nature'. Nevertheless, the title 'law of nature' has been retained for certain hypotheses which have been referred to traditionally as 'laws', such as Ohm's law, or Newton's laws of motion. Thus, while retaining the word 'law' for certain hypotheses of science, as a matter of conventional talk, it would generally be regarded as quite old-fashioned to refer to any of the recently established hypotheses of science as 'laws of nature'.

In order to classify 'laws of nature' it is indispensable to distinguish two major classes of hypotheses to one or the other of which most or possibly all (non-*ad hoc*) hypotheses of science belong. CLASS 1 hypotheses comprise well-formed as well as *idealized* empirical generalizations where the meaning of 'well-formed' will be explained in the sequel. Those *idealized* hypotheses of CLASS 1 which are referred to as 'laws of nature' will be called 'CLASS 1 laws of nature.' The other class of scientific hypotheses, which will be referred to as CLASS 2 hypotheses, consists of the starting hypotheses of hypothetico-deductive theories, such as Newtonian mechanics, electromagnetic theory, and quantum mechanics. I shall call those hypotheses of CLASS 2 which are referred to as 'laws of nature' 'CLASS 2 laws of nature.' By discussing the ontological status of hypotheses of CLASS 1 and of CLASS 2, and the interrelations of the two classes of hypotheses, I shall automatically include as special cases the ontological status of CLASS 1 and CLASS 2 laws of nature.

4.2 Empirical generalizations and Class 1 laws of nature

I pointed out that I shall refer to well-formed empirical generalizations as CLASS 1 hypotheses. At one time it was believed that one subclass of properly formulated CLASS 1 hypotheses are statements of type *S*: 'All A's are *B*', where *A*'s refer to things or to states of affairs, and *B* is a property word that refers generically to a non-defining property of *A*'s. A frequently quoted example of a statement of type *S* is 'All ravens are black'. Hempel ([8], [9]) noted that a statement of type *S* is 'logically equivalent' to either of two alternative statements *S'* and *S"*, where *S'* states that 'Everything non-*B* is non-*A*', while S" asserts that 'Everything is (non-*A*) or *B*' (where the 'or' is to be taken in the sense of the logical symbol v). Hempel soon noticed that the logical equivalence of statements *S* and *S'* led to a paradox of confirmation which many people subsequently tried to overcome (for example [10], [1], [7], [16], [17], [19], [24], [15], [20]). Indeed, if one accepts statements of type *S* as being well-formed, then a paradox readily arises. If, for example, one accepts the statement of type *S* 'All ravens are black', then this is logically equivalent to a statement of type *S'* 'All non-black things are non-ravens'. If one accepts that everything which confirms *S'* equally confirms the logically equivalent statement *S*, then the statement 'A white shoe has been found', that is an object which is a non-black non-raven, confirms the statement that 'All non-black things are non-ravens' and, hence, confirms the equivalent statements 'All ravens are black'.

One way of escaping from the paradox is to assert that statements of type *S* are not well-formed empirical generalizations. Any biologist would immediately see that this is the case, as would probably many other scientists when confronted with other examples of statements of kind *S* (for instance the statement 'All emeralds are green'). If we consider CLASS (*A*), which comprises all known *A*'s of statement *S*, and to which *A*'s discovered in future would be allocated, then individual members of CLASS (*A*) may show variability with respect to property *B* mentioned in statement *S*. For instance, multi-trait heterogeneity of classifiable organisms, such as members of human populations, or of populations of ravens, excludes their classification as a homogeneous population of individuals. Typically, human skin pigmentation shows many fine gradations among negro populations and white-skinned populations. In fact, there exist some albino negroes with white skin but negroid features and negro ancestry. Albino birds have also been observed (and photographed) and, for all we know, there could appear gene variants (alleles) for pigmentation that will lead to the development of occasional albino ravens among a raven population. Interindividual variability of properties, even if very slight in some cases, is as typical of all organismic traits as it is of the properties of inanimate systems, such as emeralds. The feather pigmentations of ravens could vary slightly among members of a population of ravens, owing to the presence in the gene pool of the population of several alleles for the pigmentation-controlling genes. Even if such alleles should not exist at present, there is always a chance that they will sooner or later be produced by random mutations. Things like emeralds might contain very minute local impurities, that could lead to colour variations at a few loci within an emerald.

Potential, or known variability of all properties of all classes of things or situations, implies that statements of type *S* are not well-formed. Accordingly Hempel's paradox of the ravens is a consequence of accepting inappropriately formed empirical generalizations as if they were well-formed. The remedy then is to give up statements of type *S* and replace them by well-formed statistically quantified empirical generalization statements. This was clearly recognized by von Wright [21] and Kyburg([14], pp. 168 ff), except that they possibly did not go far enough in taking variability into account. Thus, Kyburg ([14], p. 168) noted that

> In the case of statistical generalizations, there is no contraposition: from 'nearly all *A*'s are *B*s' one cannot infer (even probabilistically) 'nearly all non-*B*'s are non-*A*'s'. If we construe 'all ravens are black' as the limiting form of a generalization like 'nearly all ravens are black', then clearly only ravens are relevant to the generalization. If, on the other hand, we construe the same generalization as being implicitly about birds in general, then it is the limiting form of quite a different statistical generalization, namely 'nearly all birds are such that they are not nonblack ravens'. To confirm this generalization, we not only may but should examine birds of various other feathers, as well as ravens.

Yet, the possibility of observing albino ravens shows that even if we wish to confirm the generalization 'nearly all ravens are black' we must also examine within the supposed population of ravens birds whose feathers may not be black, although these birds have nevertheless recognizable (but possibly variant) defining properties of ravens. In any case, because of the possibility of feather pigmentation variability among the raven population (and the same remarks apply *mutatis mutadis* to properties of all kinds of other things and situations that are variable) one must be aware that property words like 'black', when applied to ravens (or 'green', when applied to emeralds) are only generic expressions, which do not refer to a uniquely specified property of 'blackness'. Hence, when

Kyburg uses the statement 'nearly all ravens are black', one might still ask 'In what sense are any ravens black?'

In order to deal with the variability problem one can introduce a type of well-formed empirical generalization which takes variability more explicitly into account than Kyburg's and von Wright's discussions. For instance, von Wright ([21], p. 212) asserted that 'A property which belongs to all things in a range can also be said to be *universal in that range.*' Because of variability, however, inter-individual differences of properties, however slight, may never permit rigorous talk of a 'property which belongs to all things in a range' unless we concede that the property can vary for each thing in the range considered. Accordingly I suggest the following formulation for one type of well-formed empirical generalization, which differs somewhat from that adopted by von Wright and by Kyburg, and which also avoids Hempel's paradox of the ravens. Let us consider a CLASS (A) of things or situations, and let A' be any known or discoverable member of CLASS (A). For instance, CLASS (A) could be the class of ravens and A' any particular raven. To classify any A' as a member of CLASS (A) requires 'background knowledge' (cf. Schlesinger, 1974, p. 19). The possible background knowledge could vary considerably over CLASS (A) and the accuracy of background knowledge required may depend on the specificity of the information demanded.

Suppose that for members of CLASS (A) there exists a generic PROPERTY (B), which in the case of ravens could be their feather pigmentation. The variants of PROPERTY (B) form a VARIANT CLASS (A, B), which refers specifically to members of CLASS (A). This specific reference immediately restricts the relevance (cf. [21], [14], p. 167) of the variant class to members of CLASS (A). Hence, finding variants of black shoes is not relevant to finding variants of black ravens. Let A' be any member of CLASS (A) and let B' be that member of VARIANT CLASS (A, B) which is appropriate to A', so that one can write

B' = PROPERTY VARIANT OF (B for A').

One can then, on the basis of the past observed frequency of occurrence of B', postulate a PROBABILITY (B', A) that B' will be observed when any member of CLASS (A) is inspected. Hence instead of Hempel's generalization S, discussed above, one can now pronounce a well-formed statistical empirical generalization

'For every A' of CLASS (A) there exists a B' of VARIANT CLASS (A, B) with PROBABILITY (B', A).' This formulation avoids the notion of 'nearly all' which Kyburg relied on, and makes room for the possibility that certain properties could vary considerably over a population. Moreover, the postulated probability may be subject to revision in the light of future observations.

Not all well-formed statistical empirical generalizations are of the type that I just discussed. Statistical sampling theory leaves room for different types of well-formed statistical empirical generalizations, which avoid all statements of type S (i.e. 'All A's are B's). For instance, a typical type S statement is cited by Schlesinger (1974, p. 17) 'All lead melts at 327°C'. Such a statement would at once generate a paradox analogous to that of Hempel's raven-paradox. Again, this is due to the fact that people who make such statements do not take variability of observations into account. When the melting temperature of lead is measured for different samples, slight variations of the melting temperature are observed. Even when measurements for the same sample are carried out repeatedly, observed variations of the melting temperature occur, owing to unavoidable 'observational error'. Accordingly a statistical theory of observation has been developed [12], which permits to estimate bounds for the maximum expected error deviation from some calculated mean value, in the light of past measurements. If improved measuring techniques become available, then the same statistical theory, when applied to the new measurements, may lead to changes in the calculated mean value and in the magnitude of the estimated error bounds. Hence a well-formed statistical empirical generalization states for the case under discussion that 'lead melts at T°C±Δt°C, where T°C is the best statistical average fit obtained by the statistical estimation method used (say T is a figure close to 327), and where |Δt°C| is the statistically estimated maximum error bound.' Some philosophers of science might be tempted to regard the statement just cited as an 'observation statement'. But this would be a mistake, since observations have each individually varying errors, whereas the error bound |Δt°C| and the average T°C are estimated by means of a hypotheses-laden statistical theory, so that the statement itself is a hypothesis, notwithstanding the use made of observational data in deriving the values of T°C and of |Δt°C| *via* the formulae of the statistical theory.

Some of those 'laws of nature' which form idealized empirical generalizations, and which I have earlier referred to as 'CLASS 1 laws' are related to the well-formed empirical generalizations already referred to (for instances 'lead has an average melting temperature of T°C' might be considered as an idealized empirical generalization).Other idealized empirical generalizations are based on yet another type of statistical procedure, with its underlying hypotheses (cf. [25], pp. 99 ff; [23],pp. 65-72), which, on the basis of given data, permits to fit a regression curve. This curve provides 'optimal fits' (according to some statistical fitting criterion) of the data. The regression curve is usually expressed in terms of a number of unknown parameters which are determined by a mathematical optimization procedure in such a way that the regression curve provides a 'good' fit of the data in accordance with the given criterion. Those laws of nature which can in principle be based on regression curves usually establish a relation, in terms of a regression curve, between two fitted variables and one or more fitting constants. The fitted variables represent abstractions of the observed values of variable pairs, and the regression curve represents an *idealized* statement between theoretical variables. One does not observe, however, the values of theoretical variables, but the values of empirical variables, and the values of theoretical variables and their relations provide an *idealized* representations of the observational situation. Why do people introduce theoretical variables into empirical generalizations? There are two plausible answers. First there is historical precedent. For instance, when Kepler described the orbits of planets, he wanted to represent these in 'simple', that is idealized, form, abstracted from data. The same applied to many other proponents of other CLASS 1 laws of nature. If people want to represent their relations between variables by means of simple formulae which lead to simple graphical representations of these formulae, then they obviously cannot take into account the numerous observed small (or occasionally larger) deviations from the formulae-representing curves. The fitting of relatively simple familiar curves to data antedated modern statistical curve-fitting procedures by a long time. Where possible, people just guessed suitable curves and adjusted these on an *ad hoc* basis to the given data.

The second reason for introducing theoretical variables has to do with the use of hypothetico-deductive theories. Scientific theories (if they use mathematics at all) are only concerned with theoretical variables, and not with observational fluctuations of empirical variables. Let me cite some typical (idealized) CLASS 1 'laws of nature'. Among the classics are Kepler's laws. Other examples are Newton's law of gravitation, Hooke's law for the tension of elastic strings, Ohm's law for the conduction of electricity in metals, the Wiedemann-Franz law relating thermal and electric conductivities (at ordinary temperatures) of metals, the laws of reflection and refraction of narrow pencils of light at surfaces of appropriate materials and so forth. Each one of these CLASS 1 laws applies only within stated limits, and even then, as noted, represents an *idealization*. Yet another class of statistical empirical generalizations includes some CLASS 1 laws, although the CLASS 1 laws in this case differ drastically in structure from the CLASS 1 laws discussed above. Typical examples of this further subclass of CLASS 1 laws are Mendel's laws of genetics. They represent hypotheses, which apply only to some particular species of organisms. It has been said that if the early geneticists had not been fortunate in their choices of species, and had worked with genetically more complex traits of other species, the early development of genetics might have been much delayed.

The preceding discussions lead to the following conclusions. (1) Empirical generalizations form a variety of classes, and it would be naive to assume that the characteristics of these classes can all be adequately expressed by a single master statement. (2) It appears that all well-formed empirical generalizations are, or ought to be, of a statistical nature, although this requirement is not always obvious, and has not been recognised by various philosophers of science, notably some of those who have written on Hempel's paradox of the ravens (with significant exceptions [14], [21]). Statistical considerations are required simply because of the variability of things or situations which are members of the same class. (3) Those empirical generalizations which are called 'laws of nature', and which I have here called CLASS 1 laws of nature, belong to one or the other of the numerous classes of idealized empirical generalizations. CLASS 1 laws of nature form only a minute fraction of the total number of hitherto established idealized empirical generalizations, and their being singled out under the name of 'laws of nature' is rooted in historical nomenclature.

4. 3 Hypothetico-deductive theories and Class 2 laws of nature

Let me now turn to the manner in which scientific hypotheses function within hypothetico-deductive theories (HDTs), such as Newtonian mechanics, electromagnetic theory and quantum mechanics. Such theories, once established (not infrequently in a non-systematic way), if not already deductively organized, can be reconstructed so as to assume the form of a hypothetico-deductive framework. At the hierarchically highest level of a typical reconstructed HDT are its starting hypotheses, from which in principle all possible important deductions of the theory can be made, although this is not necessarily achieved in practice. By combining various starting hypotheses in different ways, or by using them collectively, as appropriate, sometimes intermediate theorems of the HDT can be derived. Some of these intermediate theorems are occasionally called 'laws' (such as the 'law of conservation of angular momentum' in Newtonian Mechanics for the coplanar motion of a point particle about a fixed origin in the plane of motion under a force directed to that origin). Either *via* the intermediate theorems, or directly from the starting hypotheses, hierarchically lowest level deductions of the HDT can be made. These lowest level statements or formulae, as the case may be, represent idealized empirical generalizations, in that they involve theoretical variables as distinct from observed values of variables (see above). [Although quantum mechanists often refer to some of their theoretical entities as 'observables' (which are represented by operators on vectors in a Hilbert space), neither these 'observables' nor the eigenvalues of certain eigenvalue problems based on time-independent observables, are experimental variables. For instance the eigenvalues of the energy levels of the hydrogen atom calculated for the discrete spectrum from Schrödinger's equation, are theoretical variables, which must not be confused with observed values. The observed frequency of radiation emitted by a hydrogen atom, and the calculated value based on the theoretical variables can differ, because of unavoidable observational error.]

The choice of starting hypotheses of a HDT is not necessarily unique for that HDT. For instance, in quantum mechanics different exponents of the subject have chosen different starting premises, as can be seen by comparing well known textbooks, like those of Dirac [5] and Kramers [13], whose deductive systematics and starting premises differ considerably. Among some of the best known examples of starting hypotheses of a HDT are Maxwell's equations of electromagnetic theory, jointly with a class of general boundary conditions and hypothesized initial conditions (the latter being appropriate to the problem dealt with). Yet, few people, if any, would refer to Maxwell's equations as 'laws of nature', or would call Schrödinger's equation of non-relativistic quantum mechanics a 'law of nature'. There are, in fact, good reasons for refraining from this practice. For instance, Schrödinger's equation is non-relativistic, in contrast to Dirac's equation for electrons (and positrons) and in contrast to other relativistic equations for fundamental particles (cf [18])

As mentioned earlier, I shall refer to the starting hypotheses of a HDT as CLASS 2 hypotheses. I also pointed out that, because of a misplaced belief in the necessity of certain scientific hypotheses, people referred to these as 'laws of nature'. This mode of reference did not just apply to certain kinds of empirical generalizations, but was extended to the starting hypotheses of some HDTs. For instance, Newtonian mechanics of point particles in based on three starting hypotheses, which are conventionally referred to as Newton's 'laws of motion'. But in what sense are they 'laws of nature'? Newton's first 'law' of motion asserts that every body continues its state of uniform motion in a straight line unless an impressed force acts upon it. Since according to Newton's 'law' of gravitation every material system in the universe exerts a gravitational force on every other material system, a position in the universe which is absolutely force-free seems to be unattainable. Newton's first 'law' is therefore rigorously unfalsifiable, except that it is entailed by Newton's second 'law' in case of a postulated zero inertial force. The ontological status of Newton's second 'law' of motion is even more intricate. According to some philosophers of science it defines the concept of 'inertial force' **F** as being (in suitable units) equal to the rate of change of momentum **p** of a 'point particle', so that **F=dp/dt**. This definition of force *per se* would not get us anywhere, unless we combine it with appropriate definitions of constitutive forces of systems (for instance, elastic forces, gravitational forces, and so forth, as appropriate to the constitution of the 'material system' considered, subject to various idealisations). By equating constitutive and inertial forces one obtains differential equations for the motion of point particles (cf. [23] pp. 82 ff).

If Newton's second 'law' is being interpreted as a definition of inertial force, then what is 'lawful' about the second 'law'? Definitions *per se* are not necessary statements and, hence, not ordained laws. In the case of Newton's second 'law' it could be argued that if the definition of inertial force assumed a different form, if it satisfied, for instance, a different formal expression, say $\mathbf{F}=|\mathbf{p}|^2 d\mathbf{p}/dt$, then the hypothesized relation between momentum and inertial force would be different. But all that this means is that Newton's definitional relation between inertial force and momentum is based on a specific hypothesis.

Other celebrated examples of so-called 'laws of nature' are provided by the first and second 'laws' of equilibrium thermodynamics. The foundations of equilibrium thermodynamics have been beautifully expounded in classical papers by Born ([2], [3], [4]). The second 'law' together with the first 'law' (that is the 'law' of conservation of energy) permit to establish the HDT of thermodynamics. The reference to the starting hypotheses of equilibrium thermodynamics as 'laws' is a hangover from the Victorian era. Their status is no different from that of, say, Newton's equations of motion, in the sense that they are fallible initial hypotheses of HDTs. At any rate in this century, with the discovery of quantum mechanics (to replace classical mechanics when dealing with atomic phenomena and fundamental particles), with the rise and fall of the steady state theory of cosmology, and particularly since the development of molecular biology, people have fully realized that it is very unwise to refer to any hypotheses as 'laws', and the insinuated necessity of their status. For instance, the starting hypotheses of Newtonian mechanics apply only to point particles or to appropriately defined continua, provided the matter concerned does not move so fast that the special theory of relativity has to be invoked instead of classical mechanics. It is now also known that Newtonian mechanics does not apply to 'elementary particles' which require quantum mechanics for the description of their behaviour. The formulations and successes of relativistic quantum mechanics, and of relativistic quantum field theory, in cases where Newtonian mechanics did not remotely apply, showed clearly the inappropriateness of referring to the starting hypotheses of Newtonian mechanics as 'laws of motion'.

In fact, it became apparent early in this century that the alleged 'laws' of classical mechanics often led to wrong results when people tried to apply these 'laws' in order to explain phenomena of atomic physics (even when additional *ad hoc* hypotheses, such as the Bohr-Wilson-Sommerfeld quantization rules were added to classical mechanics). Thus, 'laws of nature', like all scientific hypotheses, have limitations which often only come to light after extensive applications of these 'laws'.

In view of this state of affairs contemporary scientists have become, justifiably, more cautious. For instances, it was tempting for molecular biologists to proclaim that it is a law of nature that the genetic code is universal. But wisely they desisted. Arguments favouring the temptation were summarized by Hall ([6]).

The concept of a totally universal genetic code was based initially on studies of *E. coli*, tobacco mosaic virus, and human haemoglobin. The extrapolation of this conclusion from three genetic systems to all organisms was justified by the theoretical premise that any mutation which changed the ancestral code would cause so many deleterious effects that it would have been selected against.

Here the neo-Darwinian theory of evolution was invoked as an arbiter. But, as Hall noted in his review, 'impressive as this argument may be, eukaryotic mitochondria have not heeded it.' It is known that mitochondria contain various genes, among them genes that code for mitochondrial transfer RNAs (tRNAs), which are used for translating mitochondrial messenger RNAs (mRNAs) into polypeptide chains of mitochondrially-produced proteins. It was recently discovered that, unlike tRNAs coded for by various bacterial genes and by certain nuclear genes of eukaryotic cells (so far studied), some mitochondrial tRNAs do not conform to the genetic code (see Hall's review [6]).

Notwithstanding such notable exceptions, and many others in molecular biology, there exists a general theory for many aspects of the mechanisms of protein biosynthesis. The latter theory, however, does not proclaim any 'laws of nature'. This would be utterly absurd in view of the numerous variations of details of the protein-synthesizing machinery, say between bacteria and eukaryotes (cf. [22]). For instance, ribosomes, which are instrumental in the process of 'translating' mRNAs into polypeptide chains of proteins, differ for bacteria and eukaryotes.

I must also stress that the empirical testing of a CLASS 1 hypothesis may depend on simple or complex observing machinery whose modus operandi may itself depend on one or more idealized hypotheses of CLASS 1 which are deducible from one or more HDTs. Hence, the empirical basis of establishing CLASS 1 hypotheses may explicitly or implicitly (*via* one or more HDTs) depend on the applicability of other (idealized) CLASS 1 hypotheses. This point of view has descended from Duhem (cf. [23], p. 64 for a relevant discussion) and has been championed by many philosophers of science in recent years.

4.4 Hypothetico-deductive theories and deduced idealized Class 1 hypotheses

Many scientific hypotheses depend explicitly or implicitly on certain empirical constants (such as the vacuum speed of light or Planck's constant), and there is no guarantee that these constants do not slowly change with time. Some physicists have, in fact, suggested that certain fundamental constants of nature could change slowly with the ageing of the universe. Yet, what matters for the purpose of working with HDTs that contain these possibly time-dependent 'constants' is simply that these 'constants' change so slowly with time that they can be treated as time-independent within most time-dependent differential equations that involve these 'constants'. This might not apply in cases where the differential equations used form parts of cosmological HDTs that are concerned with the ageing of the universe. A HDT, if properly formulated, must permit to deduce from its CLASS 2 hypotheses (that is, its starting hypotheses) one or more idealized CLASS 1 hypotheses. Sometimes several HDTs have to be combined. For instances, in order to deduce Ohm's law one requires Bloch's quantum theory of electronic conduction in metals, and must combine this with Fermi-Dirac formulae of statistical mechanics (cf. [26]). One may not only have to combine several HDTs in order to deduce important idealized CLASS 1 hypotheses (that is, idealized empirical generalizations) from the starting CLASS 2 hypotheses of these HDTs, but the process of making deductions is often far from straight forward. For instance, in order to be able to solve many differential equations, or other types of equations, of physics, physical chemistry and so forth, one may have to jettison certain terms of the equations on the assumption that they are 'small' compared with other terms that are being retained. In addition, even after simplifications of equations, one may have to use mathematical approximation procedures in order to deduce empirical generalizations from the basic hypotheses of theories. It may be concluded that if a supposed idealized empirical generalization deduced from a HDT does not agree approximately with observation, then the fault need not necessarily be due to an inappropriate starting hypothesis of the HDT, but could be due to excessive simplifications and inappropriate approximations used in deriving that alleged idealized empirical generalization.

As scientists succeed in inventing more accurate deduction procedures, by discoveries of, for instance, improved methods of approximation, or by applying familiar approximation methods to differential equations in which previously discarded terms are being retained, improved versions of idealized CLASS 1 hypotheses can often be derived which are in better agreement with observations than their predecessors. Another source of improved agreement between theory and observation is the discovery of novel and more sensitive experimental techniques, which reduce the extent of systematic deviations between theory and observations. For instance, the theoretical statement that the vacuum speed of light *c* is constant, does not tell us the value of *c*. Moreover, because of inevitable observational errors we could never measure the value of *c* with absolute accuracy, and, hence, could never rigorously demonstrate that *c* is actually constant. The most that we can achieve is to measure *c* with the best available techniques at any one time and try to improve on these techniques, thereby obtaining increasingly accurate mean values of *c*, without ever being able to confirm its absolute constancy. Thus, specialized theoreticians can at best claim that their theories are consistent with an estimated value of the speed of light $c \pm \Delta c$ where $|\Delta c|$ is the estimated maximum bound of observational error. The estimate is based on the theory of observation referred to earlier, and is, as emphasized before, a hypothesis-linked estimate. Hence, theoreticians can only claim that the vacuum speed of light will be found to assume a value which lies within very narrow observational error bounds such that $|\Delta c|/c \ll 1$, where the best known value of $|\Delta c|$ cannot be established *a priori* but depends on the best known available observational technique.

In order to illuminate the preceding remarks a little further let us consider the typical idealized CLASS 1 hypothesis deduced from the special theory of relativity 'for every light source the vacuum speed of emitted light is independent of the velocity of the source that emits it and equal to *c*.' I have already noted that this type of idealized statement ought to replace its claim by replacing *c* by $c \pm \Delta c$, where $|\Delta c|$, like *c*, is a theory-based quantity, the theory concerned being the statistical theory of observation. But many theoreticians are satisfied by citing simply the mean value *c*, by way of approximation, and disregard the equally theoretical term $|\Delta c|$. One attempt of testing the claimed independence of the vacuum speed of light of the source velocity was based on

> ...the study of gamma rays emitted from the decay of relativistic uncharged pions produced in very energetic nucleon-nucleon collisions. In one of these measurements the pions had an energy of 6GeV and thus a velocity very nearly that of light, so that if the velocity of the source is added to that of light, the gamma rays would have nearly twice the velocity of light, whereas if the velocities do not add the velocity is [in theoretically idealized form] just *c*. (Cited from [11]).

The observed speed of the emitted gamma rays showed that this speed approximated closely to the best known mean value of *c*, subject to stated observational error bounds. The difference between the estimated vacuum speed of light (within observational error bounds) and the estimated speed of the emitted gamma rays was non-random and of a small persistent order of magnitude in accordance with the sensitivity of the observing technique used.

Recent more sensitive experiments (see [11]) which relied on regular pulsing X-ray sources of binary stars, showed a systematic deviation of the estimated speed of emitted X-rays from the estimated vacuum speed of light, which was about 10^{-5} times that obtained with the pion method. This example illustrates that improved methods of observation, although never providing any exact value of any observed quantity, can improve drastically on previous estimates of (exactly unknowable) physical constants.

The upshot of this paper is the contention that so-called 'laws of nature' do not differ in classificatory respects from a host of classifiable hypotheses of science which are not referred to as 'laws of nature'. To refer to some statement as a 'law' is only in keeping with conventional scientific language usage, but does not confer on the statement a special status.

University of Newcastle-upon-Tyne

Bibliography

[1] Ayer, A.J., *Probability and Evidence*, Macmillan, London, 1972.

[2] Born, M., 'Kritische Betrachtungen zur traditionellen Darstellung der Thermodynamik', *Physikalische Zeitschrift, 22* (1921), *218-224*.

[3] Born, M., 'Kritische Betrachtungen zur traditionellen Darstellung der Thermodynamik', *Physikalische Zeitschrift,* 22 (1921), 249-254.

[4] Born, M., 'Kritische Betrachtungen zur traditionellen Darstellung der Thermodynamik', *Physikalische Zeitschrift,* 22 (1921), 282-286.

[5] Dirac, P. A., *The Principles of Quantum Mechanics* (4th ed.), Clarendon Press, Oxford, 1958.

[6] Hall, B. D. 'Mitochondria spring surprises', *Nature* 282 (1979), 129-130.

[7] Harré , R., *The Principles of Scientific Thinking,* Macmillan, London, 1970.

[8] Hempel, C. G., 'Studies in the Logic of Confirmation I', *Mind* 54 (1945), 1-26.

[9] Hempel, C. G., 'Studies in the Logic of Confirmation II', *Mind* 54 (1945), 97-121.

[10] Hempel, C. G., ' A Note on the Paradoxes of Confirmation', *Mind* 55 (1946), 79-82.

[11] Hodgson, P. E., 'Speed of Light and Relativity', *Nature* 27 (1978), p. 13.

[12] Janossy, L., *Theory and Practice of the Evaluation of Measurements,* Clarendon Press, Oxford, 1965.

[13] Kramers, H. A., *Quantum Mechanics,* North-Holland, Amsterdam, 1957.

[14] Kyburg, H. E., *Probability and Inductive Logic,* Macmillan, London, 1970.

[15] Mackie, J. L.,'The Paradoxes of Confirmation', *British Journal for the Philosophy of Science* 13 (1963), 265-277.

[16] Salmon, W. C., *The Foundations of Scientific Inference,* University of Pittsburg Press, Pittsburg, 1969.

[17] Schlesinger, G., *Confirmation and Confirmability,* Clarenden Press, Oxford, 1974.

[18] Schweber, S. S., *Relativistic Quantum Field Theory,* Harper and Row, New York, 1964.

[19] Smokler, H., 'The Equivalence Condition', *American Philosophical Quarterly* 4 (1967), 300-307.

[20] Swinburne, R. G., 'The Paradoxes of Confirmation - A Survey', *American Philosophical Quarterly* 8 (1971), 318-329.

[21] Von Wright, G. H., 'The Paradoxes of Confirmation', in J. Hintikka and P. Suppes, eds. *Aspects of Inductive Logic*, North-Holland, Amsterdam, 1966, pp. 208-218.

[22] Wassermann, G. D., *Molecular Control of Cell Differentiation and Morphogenesis*, Marcel Dekker, New York, 1972.

[23] Wassermann, G. D., *Brains and Reasoning*, Macmillan, London, 1974.

[24] Watkins, J. W. N., 'Confirmation, the Paradoxes and Positivism', in M. Bunge, ed. *The Critical Approach to Science and Philosophy*, The Free Press of Glencoe, 1964.

[25] Weatherburn, C. E., *Mathematical Statistics*, University Press, Cambridge, 1952.

[26] Wilson, A. H., *The Theory of Metals*, University Press, Cambridge, 1936.

5 Biological aspects of the philosophy of mind

5.1 The issue

Concerning the nature of philosophy, Ayer wrote [1]

> The thread which is emerging is that philosophy has to do with criteria. It is concerned with the standards which govern our use of concepts, our assessments of conduct, our methods of reasoning, our evaluations of evidence. One thing which it may do is to bring to light the criteria which we actually employ; another to adjudicate if they are found to conflict; another, perhaps, to criticize them and to find better substitutes for them.

Many philosophers believe that in the pursuit of these aims problems of philosophy ought to be rigorously separated from problems of science, except, possibly, within the philosophy of science. They deny that the sciences can contribute to the resolution of philosophical problems. Among these problems are (1) the search for criteria of truth, which has resulted in several 'theories' of truth,[2] and (2) requests for classificatory analyses of our use of concepts, so as to avoid, for instance, paradoxes created by the faulty use of concepts. In this paper I wish to suggest that, contrary to much prevalent opinion, science, in the guise of neuropsychology and related biological topics, can make a definite contribution to the resolution of deep philosophical problems.

The construction of theories of truth *per se* does not tell us which things in the world are true and which are not, and the urge for the attainment of truth is often linked with a longing for certainty. For instance, long ago David Hume[3] asserted that mathematical statements 'are discoverable by the mere operation of thought, without dependence on what is anywhere in the universe...' and that even if 'there never were a circle or triangle in nature, the truths demonstrated by Euclid would for ever retain their validity and certainty and evidence.' Kant held similar views.[4]

In our century Bertrand Russell[5] asserted more generally, but similarly, that 'a philosophical proposition must be such as can be neither proved nor disproved by empirical evidence', suggesting implicitly that philosophical propositions ought to be purely 'a priori'. Hume and Kant might, very likely, have conceded that mathematical axioms, which *per se* are unprovable (otherwise they would not be axioms), play a prominent part in mathematics, but they would have claimed that metamathematics, metalogic and philosophy should not depend on empirical hypotheses. This view was also shared by Wittgenstein, who thought that our criteria for the use of concepts should only rely on analyses of concepts as we find them embedded in linguistic contexts, and no hypotheses about languages should be invoked in philosophical analyses of language usage. Seen from this point of view the work of Chomsky and many other theorists of language structure would seem to be purely scientific.

In the domain of empirical propositions it has long been realized that Humean scepticism rules out certainties. Scientific theorizing depends on very uncertain, unprovable hypotheses and theories, which at best can be confirmed, and whose replacement rate, because of falsification by evidence or appearance of better theories,[6] is often impressive. Accordingly the truth of scientific theories cannot ever be asserted, with absolute confidence, by any truth theory. What applies to science applies on a less dramatic scale to the propositions of mathematics and logic. Hume, if he lived now, might have applied his sceptical attitude as much to mathematical as to empirical propositions, instead of proclaiming certainty of the former. It has been found that questions concerning the decidability of the consistency of certain axiomatic systems are sometimes answered in the sense that the consistency of these systems is undecidable. Hence, we must remain uncertain about the consistency of various axioms within certain mathematical systems. The prevalent uncertainties about scientific theories, and about purely mathematical systems, are paralleled by similar uncertainties about the results of linguistic analyses, for the simple reason that one can never conceivably exhaust *a priori* all possible contexts in which language expressions can occur on future occasions. To complicate matters, Quine, while discussing analyticity,[7] considered that analyticity may be related by empirical hypotheses to language learning, thus blurring the border between analytical and synthetical propositions.

The uncertainties and blurrings just referred to can be reduced in significance, I think, if we adopt Popper's theory of knowledge, which 'is very simply that all we ever have at the very best is a set of hypotheses [or axioms in the case of mathematics]; however well authenticated, they are never more than that. The best established piece of knowledge may have to be modified or given up.'[8] Unfortunately Popper's views on epistemology have been slow to penetrate to certain philosophers. If knowledge relies in many cases either on hypotheses or on unprovable axioms, and must always remain incomplete as illustrated by language usage analyses, then, surely, *philosophical* knowledge could hardly aspire to certainties and hope to remain hypotheses-free, simply because some philosophers say so. I propose to defend the view that, notwithstanding the uncertainties of scientific theories, their introduction into philosophical arguments may substantially assist in solving philosophical problems.

For a start let us look at Hume's claim concerning the eternal certainty of Euclidean geometry 'irrespective of what is anywhere in the universe'. His claim is metaphysical (in Popper's sense) since it could only be tested by removing all material systems from the universe. Worse for Hume, neuropsychologists claim that the production and understanding of thoughts has a biological basis vested in human brains and their evolution. Since brains 'exist in the universe' (at least as far as philosophical realists are concerned), Hume's conclusion seems unacceptable to neuropsychologists, although it might appeal to pure idealists. Admittedly neuropsychology and its theories are as fallible as the thoughts produced by brains. Yet, it is an often disregarded truism that all statements, whether classified as analytical or empirical, as valuative or normative and so forth, are brain produced (or produced by artificial intelligences [AIs]) and that such statements can only be understood with the help of very specific material systems namely human brains. Books in libraries become only 'meaningful' when operated on by human brains, or conceivably by AIs at some future date. The role played by brains in 'human understanding', particularly in the possible solution of philosophical problems, is far from trivial as soon as we enquire how precisely brains transact various classes of cognitive functions.

For instance, when people inspect philosophical books or papers in libraries they require appropriate brain representations of retinal images to start with. The representations could be encoded in pictorial form, thereby preserving topographical and topological relations of retinal images, or they could be expressed, as some theorists believe, in purely symbolic non-pictorial codes.[9] In fact, in 'creative thinking' brains can generate new image representations, of things never perceived, and can sometimes translate these, with the help of motor systems, into paintings. This suggests also that, *pace* Hume, knowledge of figures such as circles and triangles depends on brain representations, and, hence, on 'what there exists in nature'. In fact, people could not recognize triangles and circles for what they are if there did not exist cerebral memory traces (engrams) which represent these configurations. Without engrams pattern recognition would be impossible. I have suggested, in terms of a neuropsychological theory, that apart from engrams of images there exist engrams for concepts some of which can represent images in token forms, and that there exist engrams which are hierarchically organized and which can represent hierarchies of concepts, where many of these concepts (such as 'kingdom', 'household goods', 'Abelian group') need not, and do not, correspond to images. In addition I have postulated that these hierarchies of concept-representing engram systems can form or become linked into intermodal or intramodal 'concept-representing association complexes'.[10] At the level of brain representations there need to be no essential difference in the types of mechanisms which represent concepts of abstract notions and those which represent concepts of concrete entities, except that the brain mechanisms used in representing hierarchically higher order, and hence more abstract, notions may involve more complex 'association complexes' of the same type as are used in representing simpler concepts.

When Russell argues that philosophical propositions should be independent of empirical evidence (see above), we may ask 'What is to be regarded as empirical evidence?' If 'empirical evidence' is provided by our acquaintance with things in the world, then this acquaintance requires the formation of engrams of things that we have perceived in the world, and engrams of the related concepts which we have formed. Thus engrams could, and almost certainly do, represent 'empirical evidence' in a physical form which differs from the things which the engrams represent, but nevertheless engrams ought to be ranked as 'empirical evidence'. Since brains can generate new images and their corresponding concepts in 'creative thinking', and can also create representations of novel concepts which need not be image-related, the 'creatively produced' engrams of either images or concepts or both, can be classified as much as 'empirical evidence' as engram systems which refer to things in the 'external world'. What counts as 'empirical evidence' in terms of brain representations lies in the eye of the beholder. An abstract concept acquired by reading a textbook need not differ with regard to its basic mode of brain representation from that of an abstract concept invented by the reader. I conclude that if we admit neuropsychology to deal with some of the problematics of philosophy, then Russell's definition of a 'philosophical proposition' appears arbitrary, vague and unacceptable. He apparently failed to realize that abstract arguments of mathematics may depend as much on 'empirical evidence' in the form of brain representations as scientific arguments, or everyday discussions. Thus, neuropsychological theories can throw important light on philosophical problems. I propose to reinforce this conclusion in the rest of this paper.

5.2 Neuropsychological answers to some problems of the philosophy of mind

I shall consider a few typical problems in the philosophy of mind. Let me first try to provide a neuropsychological answer to the philosophical question 'Why is the true answer to a problem the answer it is?' I shall only sketch my answer at surface level, since possible and sufficient machinery at the neuropsychological level has been postulated and explained in great detail elsewhere.[11] An 'answer' to a problem of any kind is something behavioural, given in speech, writing or other physical symbolic representations such as sign language. The relevant behaviour that leads to the answer could depend on the creation by brains of programmes and sequences of programmes, the programmes consisting of sequences of representations of images or of concepts as appropriate, and the items of these sequences could, for instance, be activated by means of fast-acting biological chronometers in precise serial order. The representations could rely on already existing engrams or on the formation of new engrams by means of 'creatively functioning' engram-generating processes, that is, processes that do not depend on current perceptual input.

Perhaps one of the simplest types of 'problems' demands the recognition of an object or of a situation by naming a presented object or situation. I suggested recently that this could mean that all existing engrams are in parallel confronted with the same set of representations of the object presented. The representation set could consist of the representation of a particular variant, say a size variant or a shape variant, of the object. By providing mechanisms which permit different size-, shape-, orientation (etc.) editions of the same primary object representation to be synthesized, all available image-representing engrams could in rapid sequence be confronted with all possible variants of the configuration of the presented object that the brain can synthesize in representational form by means of standard mechanisms,[12] until one of the variants (of a familiar configuration) matches one of the existing engram representations of that configuration. Parallel representations of the same configuration to all configuration-representing existing engrams makes for great economy in search procedures, so that appropriate matching could be very rapid. When matching between a synthesized variant representation of an image pattern and an engram of the pattern takes place, this event could be signalled to a specific structure which represents the engram of the concept that corresponds to the matched engram of the pattern. This pattern-concept-representing structure, in turn, could send a specific signal to a particular hierarchy of language-concept-representing 'association complexes' (see above). These, association-complexes, in turn, when activated, could send signals to systems which represent engrams of hierarchies of motor acts of the kind that permit to verbalize the language concepts, and so forth. In this way, recognition of an object could lead to its appropriate verbal conceptualization which, in turn, could lead to verbal descriptions of the object and its properties.

In the example just described, we see that if somebody names and describes a pattern correctly then 'the true answer is the answer it is' because the brain's representational machinery operates sequentially and in parallel, by means of a host of representations, in such a fashion that it provides an answer or statement which corresponds in linguistic terms to the object or situation which has been recognized. Thus the neuropsychological account helps to clarify in terms of a scientific theory why and how the 'correspondence theory' of truth[13] could work in the case under consideration. The neuropsychological theory invoked shows, incidentally, that, because of its reliance on language engrams, the validity of the statement, that is, its 'truth by correspondence', depends on consensus. It demands that members of a given society have learned to use certain linguistic expressions and have learned to associate these with certain perceptual representations of specific objects or situations. This requires learning processes of a kind that help to establish many of the engram systems for images and concepts which I have invoked as given in the preceding account.

I am only claiming that neuropsychological theories can provide possible, but not ***unique***, illuminations of the 'correspondence theory of truth' for the example considered. Different neuropsychological theories could provide different kinds of illuminations of the philosophical problem raised. Uncertain as these theory-based illuminations are, they provide coherent explanations, and this cannot be said of the most sophisticated common sense accounts since they explain nothing. To say that an apple *is* an apple because we *call* it an apple is no more than a statement of a hidden problem, which neuropsychological theories try to resolve in one way or another. Essentially common-sense answers state only that languages function so as to express what they express, and that we can analyze what they express. But this provides hardly any understanding of the real puzzle why languages can express thoughts or perceived states of affairs. The 'sophisticated' common sense answer turns out to be a platitude, and not an insight. The question is 'Do philosophers only want profound-sounding platitudes or do they in addition care for insights based on uncertain theories?'. Scientists, and philosophers like Popper, are satisfied with very uncertain theorietical answers which may require revision. Why should other philosophers wish to aim at certainty in a world where certainty cannot be attained? Admittedly, neuropsychology at theory level is reductivist, but so are a host of scientific theories, for example, theories dealing with solid state physics in terms of postulated atomic constructs. Perhaps, saying that an apple is an apple because we call it an apple saves one much time in acquiring knowledge about neurobiology and complex neuropsychological theories. It might look provocative, but I believe that deep insights cannot be bought by cheap evasions of problems.

The above illustration of 'why the true answer is the answer it is' concerns only a very simple type of 'problem solving', namely naming a presented object or situation correctly. Engram representations of complex concepts and images can solve vastly more complex problems which may involve intricate trial-and-error search procedures. The solution of any one problem may rely on essentially similar classes of mechanisms. There may be many different classes of problems which utilize the same general types of mechanisms in problem-specific ways. Accordingly no universally valid answer can be given in terms of neuropsychological theories why 'the true solution of a problem is the solution it is'. For each class of problems we might provide an appropriate neuropsychological explanation. But such issues arise in similar ways throughout science. For instance, Maxwell's electro-magnetic theory can be applied to a host of different systems which depend on different classes of mechanisms to which Maxwell's theory applies. Yet, Maxwell's theory provides no universally utilizable answers for all the classes of mechanisms to which it applies. It has to be used in specific ways for particular members of each appropriate class of mechanisms. This tedious situation is inherent in science, and has resulted in millions of experimental and theoretical papers which investigate the mechanisms of myriads of different systems. There are no good reasons why philosophers should be able to solve their problems by universally applicable answers. Analysts of language usage have certainly realized long ago that there are myriads of 'problems'. But their problems arise at the common sense level of analysis, and I suggest that other types of science-based analyses of philosophical problems are possible which also require much piecemeal work.

Next let me turn to the philosophical problem 'Why do we perceive the world as we do?' At the purely behaviourist level when somebody who is shown an apple, and, if asked what it is, answers 'it is an apple' - the problem has already been dealt with in terms of the preceding neuropsychological explanations. I have sketched how, in neuropsychological terms, a presented apple can lead to the statement concerning the identity of the presented object. But if we ask why we normally always have *specific* private experiences when we see an apple, then we can bring in epiphenomenalism, which is defensible[14] if suitably expounded. If epiphenomenalism is to do a proper job, then it must assert more than the vague statement that particular brain states are accompanied by particular private experiences as epiphenomena. A more precise formulation is possible. We can, for instance, assume for one class of neuropsychological models[15] that they provide in large numbers of structurally similar copies primary physical representations (as distinct from engrams), of say, visually presented objects (and similar remarks could apply to other sensory modalities *mutatis mutandis*), and that the same applies to multiple representations of engrams.[16] If any particular one of these structure copies, which represent in primary (non-engram) form a specific object, is appropriately activated, then a specific epiphenomenal experience could turn up, which corresponds to the state of the activated representative structure while it represents the particular object. Accordingly the epiphenomenon is object-specific.

Similar remarks could apply to reactivated engrams during recognition processes or during memory retrieval. Specific activated engrams are accompanied by specific epiphenomena representing specific retrieval of remembered things or of concepts, depending on what the engrams concerned represent. The way in which specific engrams could be formed and stabilized and represented in multiple editions was suggested elsewhere.[17] It must be stressed that although, in the terms stated, epiphenomenalism may make sense it remains, nevertheless, a metaphysical hypothesis, since we cannot detect unambiguously what kinds of brain structures represent private experiences, and since we cannot publicly observe these experiences. There could be many potential candidate structures, and all we can show is that certain candidate structures would suffice. By contrast, it seems more profitable to link language (and other) behaviour *via* neuropsychological theories with neurobiological structures and their dynamics which are in principle testable. But nobody should be under any illusion that at present adequate tests are feasible, because of the huge number of components involved in nervous transactions (some 10^{12} neurons for the human bran, and with possibly 30,000 synapses - one percent of these being operational - for the average [rat] cortical neuron, and so forth).

Faced with this situation all we can do at present is to suggest candidate neuropsychological theories and elaborate these in detail, and see how they could explain in detail a host of phenomena studied by experimental psychologists in neurobiological terms. I have mentioned that the type of neuropsychological theory that I have recently developed in detail could explain how perceptual representations of things and their concepts could be translated into linguistic representations. Yet language terms can hardly be expected to be in one-one correspondence with the things that activate our sensory surfaces. Language can only be expected to provide crude symbolic tokens of much more detailed and explicit representations provided by perceptual mechanisms and their presumptive epiphenomena. Language is a poor substitute for things perceived. I must also stress that many neuropsychological theories have been propounded, and most of these are seriously deficient when gauged against performance requirements.[18]

5.3 Common sense analyses and neuropsychological insights

I am not arguing that common sense analyses should be dispensed with in favour of neuropsychological explanations, rather, that the two kinds of approaches should be complementary. Both types of enquiries have their serious limitations, as noted. For instance, common sense linguistic analysis is particularly weak when confronted with the problems of the philosophy of the mind, which are intimately related to issues of neuropsychology. Thus, when we wish to illuminate what we mean by the concept of 'concept', then one way of illuminating this is to discuss how concepts could be represented and *function* within neuropsychological theories. Linguistic common sense analysis does not offer any adequate substitute for this. In fact, if we enquire how the word 'concept' is used in language, then we ought to consider not simply everyday languages, but also scientific languages, such as the discourse of neuropsychology, which deal with the concept of 'concept'. Similar remarks apply to many other concepts that concern the philosophy of mind. To consider concepts as hypothetical constructs of neuropsychological theories is no more extraordinary than to regard atoms and molecules as hypothetical constructs of atomic physics. There exists an obvious difference of goals between people who aspire to reductionist explanations of the *modus operandi* and of the structural basis of concepts in terms of neuropsychology, and those who are content with mere linguistic indicators at the common sense level of how concepts are used. Many AI designers and neuropsychologists would question whether common sense analyses have given us an adequate understanding of what 'mentality' amounts to.

Consider for instance the concept 'intelligence'. It is a commonplace to assert that 'intelligence' is 'measured' by a single parameter quantity called the IQ. This type of analysis, if you wish to call it that, evades the issues, since it provides no information of the structural and functional competences that are involved in the brain *processes* of problem solving, or of pattern recognition or of numerous other intellectual tasks. The one-parametric approach to intelligence closely resembles the one-parametric approach to inheritance, based on a measure called 'heritability'. Feldman and Lewontin severely criticized[19] such single (or few) parameter measures since they provide no information about the components of the machinery and the processes enacted by the machinery which represents such things as the development of nervous systems and the influence of genes on that machinery or such things as intelligence in terms of recognition processes, problem solving and so forth.

To argue about processes which we term 'intelligent' at the level of IQs, or in terms of common sense linguistic analyses, is, to use Feldman and Lewontin's related conclusions (which refer to 'heritability'), 'nearly equivalent to no information at all' for any serious problem of intelligence. One does not really understand properly what one is talking about in relation to such things as genetics and intelligence until one has thoroughly understood the classes of mechanisms that are involved. (One does not only want to know what genes there are but also what they do and *how* they help produce phenotypes). IQ, or language analyses of concepts relating to 'mentality', provide only 'poor' and, worse, misleading substitutes for pictures of the machinery (as Feldman and Lewontin remarked in relation to 'heritability'). In fact, in connection with their assessment of 'heritability' Feldman and Lewontin concluded that 'there is a vast loss of information in going from a complex machine to a few descriptive parameters.' Much that is relevant at the process level is lost at the level of linguistic common sense analyses of complex intellectual phenomena. If the essence of mental behaviour can only be grasped by understanding the workings of the machinery that produces it, then no amount of witty language analysis of the concept of 'Mind' a la Ryle,[20] could provide a proper understanding of what we mean by cognitive processes or activities.

A fashionable approach to 'intelligence' and, more generally to concepts of cognitive psychology, relies on designing artificial intelligence (AI) models,[21] usually expressed as computer programs. AI programs depend, as a rule, on the 'intelligence' of human programmers who produce these programs. By contrast, human brains are self-programming, and need not work like computers at all. By means of developmentally provided neurobiological equipment brains can themselves synthesize and encode the procedures which they utilize. Apart from this, it cannot be concluded that the procedures utilized and synthesized by brains, and the modes of representations of AI procedures used by computers resemble each other in any way.

Returning to common sense analyses, I must stress another reason for its ultimate incompleteness. Analyses of concepts in terms of language usage depend invariably on prior knowledge of other concepts that form the background against which the use of a concept is evaluated. This applies, for instance, when we consult dictionaries which elucidate one concept in terms of others. These procedures will ultimately either be circular or result in a residue of unanalyzed concepts. In order to 'know' a concept and to use it appropriately the concept has ultimately to be represented by brains in terms of appropriate engrams. This leads back to neuropsychological theories about the modes of representations of concepts by brains. It might be objected that knowing that each concept is represented by a specific engram does not help us in analyzing that concept. We do not learn to understand words by knowing that there are engrams for these words. We must also remember that a child does not learn the meaning of words either by means of dictionaries. He learns to associate words with things and situations and learns how other people use words. But then philosophers do not only want to know simply the meaning of words, but they also want to know why certain words refer to certain things, and why this is possible, and this is where neuropsychology comes in.

We also want to know why, without linguistic analyses, ordinary people can usually understand each other's talk. This, I suggest, depends ultimately on similarities of the wiring-in of human nervous networks in species-specific ways, and it also depends on the provision of species-specific mechanisms for the presentations of such things as images and engrams, notably engrams associated with images and concepts. I conclude that whereas the common sense philosophers may be primarily interested in answering what words symbolize, philosophers of mind ought also to be interested in knowing why words symbolize what they symbolize. Elsewhere[22] I advocated that the relationship between philosophy (particularly the philosophy of mind) and neuropsychology could profitably be based on a feedback cycle which involves successive approximations. Improvements in neuropsychological theories should enable philosophers to deepen their understanding of the relations between language, perception and thinking. Improved philosophical insights based on better neuropsychology, in turn, should allow philosophers to enlighten neuropsychologists more firmly about the epistemological foundations of neuropsychological theorizing, for instance, what theorists mean by hypothesis formation, by 'empirical evidence', by deduction, how philosophy can cope with the problems of scientific induction, and so forth.

The University of Newcastle Upon Tyne

Notes

1 A. J. Ayer, *The Central Questions of Philosophy* (London: Weidenfeld and Nicolson, 1973), p. 2.
2 Cf. K. Lehrer, *Knowledge* (Oxford: Clarendon Press, 1974), Chapter 2.
3 D. Hume, *Enquiries* (edited by L. A. Selby-Biggs, 2nd ed.) (Oxford: Clarendon Press, 1902), p. 25.
4 Cf. G. D. Wassermann, *Brains and Reasoning* (London: Macmillan, 1974), p. 14.
5 B. Russell, *Mysticism and Logic* (Harmondsworth: Penguin, 1953), p. 107.
6 I. Lakatos, "Falsification and the Methodology of Scientific Research Programmes" in *Criticisms and the Growth of Knowledge* (I. Lakatos and A. Musgrave eds.) (Cambridge: Univ. Press, 1970).
7 Cf. W. V. Quine, *The Roots of Reference* (La Salle Ill.: Open Court, 1973), p. 79

8 J. O. Wisdom in *The Critical Approach to Science and Philosophy* (M. Bunge ed.) (London: Collier-Macmillan, 1964), p. 116.

9 For detailed discussions of possible codes see G. D. Wassermann, *Neurobiological Theory of Psychological Phenomena* (London: Macmillan and Baltimore: University Park Press, 1978), Chaps. 1 and 3.

10 Op. cit., note 9, Chap. 5.

11 Op. cit., note 9.

12 Op. cit., note 9.

13 See op. cit., note 2

14 G.D. Wassermann, 'Reply to Popper's Attack on Epiphenomenalism', *Mind* vol. 88 (1979) p. 572

15 Op. cit., note 9.

16 Op. cit., note 9 and K.S. Lashley, 'In Search of the Engram' in *Symposia of the Society for Experimental Biology,* Vol. 4 (1950) p. 454.

17 Op. cit., note 9.

18 Op. cit., note 9, Chap. 6.

19 M.W. Feldman and R.C. Lewontin, 'The Heritability Hang-up', *Science,* Vol. 190 (1975), p. 1163.

20 Cf. G. Ryle, *The Concept of Mind* (London: Hutchinson's University Library, 1949).

21 M. Boden, *Artificial Intelligence and Natural Man* (Hassocks, Sussex: The Harvester Press, 1977).

22 Op. cit., note 4.

6 Reply to Popper's attack on epiphenomenalism

Until recently Popper did not distinguish between the use of the concept 'natural selection' as it appears (1) in falsifiable population genetic theories and (2) in unfalsifiable metaphysical evolutionary speculations. In his 'First Darwin Lecture' Popper[1] has, as he put it, 'recanted' his previously held views concerning the alleged universal untestability of natural selection, and some of his new attitudes seem to harmonise with my own differently argued conclusions.[2] Popper's First Darwin Lecture also attacked Huxley's defence of epiphenomenalism, the attack being based on a supposed clash between natural selection and epiphenomenalism. A similar argument appeared also in a recent book by Popper and Eccles.[3] I shall argue that Popper's attack on epiphenomenalism which appears in his book with Eccles seems to me to be unsatisfactory.

Popper (ibid. note 3, pp. 72-74) noted that

> Huxley[4] puts his epiphenomenalism very well: 'Consciousness...would appear to be related to the mechanism of [the] body, simply as a...[side] product of its working, and to be as completely without any power of modifying that working as the [sound of a] steam-whistle which accompanies the work of a locomotive...is without influence upon its machinery.

Following this, Popper claimed that Huxley's 'epiphenomenalism clashes with the Darwinian point of view'. In fact Popper believes that 'The Darwinian view must be this: consciousness and more generally the mental processes are to be regarded (and, if possible, to be explained) as the product of evolution by natural selection'. In line with this perfectly reasonable conclusion Popper argued that

> The Darwinian view is needed, especially, for understanding intellectual mental processes. Intelligent actions are actions adapted to foreseeable events. They are based upon foresight, upon expectations; as a rule, upon

short term *and* long term expectation, and upon the comparison of the expected results of several possible moves and countermoves. Here *preference* comes in, and with it, the making of decisions, many of which have an instinctual basis...

Popper believes that selection pressures account for the coherence in the cultural world. 'For example, the evolution of language can be explained, it seems, only if we assume that even a primitive language can be helpful in the struggle for life, and that the emergence of language has a feedback effect: linguistic capabilities are competing; they are being selected for their biological effects; which lead to higher levels in the evolution of language.' Although it seems biologically far from obvious that evolution of anything, including language, must be 'progressive' I shall not digress on this particular issue.

His belief in neo-Cartesian interaction dualism apparently inspired Popper's attempt to discredit epiphenomenalism. He maintained, correctly, that 'the epiphenomenalist must insist [that the world of private experiences] is indeed irrelevant; that only physical processes matter: If a man reads a book, the decisive thing is not that it influences his opinions, and provides him with information. These are all irrelevant epiphenomena. What matters is solely the change in his brain structure that affects his disposition to act. These dispositions are indeed, the epiphenomenalist will say, of the greatest importance for survival...'. The epiphenomenalist, Popper goes on to say, insists that if we read something then this may change our brain structure (e.g. by forming new engrams for short-term memories), but the subjective experiences that may occur when reading are epiphenomena of the physical brain states and physical changes of brain states and structures. In preparation for his main argument Popper notes that epiphenomenalists admit the existence of a world of conscious experiences, which are for them epiphenomena of brain states, but that epiphenomenalists deny conscious states any biological function. From this Popper tried to deduce that epiphenomenalism could not explain the evolution of conscious experiences in Darwinian terms. Hence, in Popper's opinion, epiphenomenalists are forced to deny 'what [according to Popper] is plainly a most important fact', namely the tremendous impact of this evolution of conscious experiences (and of the world of culture) on the world of physical things.

I believe that there are some serious flaws in Popper's main arguments, and that he fails to discredit epiphenomenalism. To start with, natural selection acts on phenotypes and, *via* these, on genotypes. Genetic systems, jointly with environmental factors, could control developmental processes and, hence, could suffice to determine the wiring-in of nervous systems.[5,6] In particular this could help to explain the determination of the neural basis of so-called 'innate' behaviour of animals, notably the capacity of many animals to respond to sign-stimuli of other animals of their own species or of predators (cf. Lorenz's beautiful account of the mating behaviour of the Salticid spider, quoted on p. 24 of op. cit. in note 6). What matters, therefore, is the sufficiency of molecular genetic control systems, together with environmental factors, for the wiring-in of nervous systems. Popper has not proved, by his arguments, that molecular genetic control mechanisms of developmental processes plus environmental physical factors must necessarily be insufficient for laying down all of those systems that could account for the physical aspects of mental processes, including all forms of cognitive processes. (Learning processes could modify the nervous system which is wired-in under genetic control; cf. op. cit. in note 6). Since genetic control mechanisms, jointly with environmental factors, could be sufficient for the wiring-in and modification of nervous systems, one does not have to invoke subjective experiences as factors involved in establishing or modifying nervous systems. Hence, up to this point the epiphenomenalist doctrine remains viable. It must be stressed that only the sufficiency of this doctrine could be established on theoretical grounds, but that the doctrine itself is as untestable as interaction-dualism, and that the latter, like epiphenomenalism, is a metaphysical doctrine.[7]

Since natural selection acts on genetic systems (*via* phenotypes; cf. op. cit. in note 2), it is relevant that by changing those genetic subsystems which are involved in the control of the wiring-in of nervous systems (cf. op. cit. in note 6), this wiring-in could be changed if, for instance, the genes concerned experience mutational changes. This could lead to changes in phenotypical behaviour of animals with changed nervous systems, and lead to appropriate natural selections. But what is then being selected for, according to the epiphenomenalist, are genes that control the wiring-in of nervous systems, and not conscious experiences, since it is consistent with epiphenomenalism that conscious experiences, unlike genetic systems, have no effects on phenotypes. Even if genetically altered, naturally selected, nervous systems produce different epiphenomena, then this would still be fully consistent with epiphenomenalism. The fact that epiphenomenalism cannot explain in Darwinian terms the evolution of consciousness in detail does not speak against epiphenomenalism, since epiphenomenalism does not try to provide a detailed theory of which particular epiphenomena correspond to particular brain states. An epiphenomenalist could simply argue that the types of epiphenomena that nervous systems can produce depend on the complexity of these nervous systems. Hence, as nervous systems evolved to greater complexity, by purely physical evolution, this permitted more complex functions of the more highly evolved nervous systems, and, hencc, quite possibly, allowed genesis of more intricate epiphenomena. Accordingly the progressive evolution of higher consciousness is fully consistent with a purely physical evolution of nervous systems, based on the natural selection of genotypes *via* physically produced phenotypes, and there is no need to invoke conscious states as modifiers of phenotypes.

It would be difficult to defend the view that the evolution of nervous systems should depend on consciousness, rather than the opposite view, i.e. that the evolution of the capacity to form more complex epiphenomena depends on the prior evolution of more complex nervous systems. After all, innumerable somatic structures have evolved without the assistance of consciousness. There exists no evidence that livers, hearts, and other non-nervous somatic systems of mammals and other organisms have evolved with the help of conscious experiences. In the case of plants there exist no nervous systems, and no evidence that their evolution was governed by consciousness. I conclude that the epiphenomenalist could claim that evolution has helped to produce better nervous systems and has thereby probably provided a better capacity to form more complex epiphenomena, but that conscious experiences are quite redundant for the evolution of systems capable of forming conscious states.

Popper argues as if subjective experiences were parts of the phenotypic features on which natural selection acts. Yet the epiphenomenalist could simply retort that such selection as occurs takes place in relation to the systems that produce the epiphenomena themselves. Subjective experiences are not known to the environment, only behaviour can be known, and natural selection can act on behaviour, which is not the same as acting on (epiphenomenally produced) subjective experiences. The influences which nervous systems have on the external world, including on our cultural life, are *via* behaviour and not *via* private experiences. Popper's argument appears to substitute private experiences in place of publicly observable behaviour. No epiphenomenalist would deny that the behaviour of organisms can have selective effects. But this is not the same as claiming that private experiences cause behaviour modes that could have selective effects. Epiphenomenalists simply assert that behaviour is caused by nervous systems jointly with other somatic systems and environmental physical factors. They claim that nervous systems while being instrumental in producing behaviour may in some (but not necessarily all) cases also produce epiphenomena. And they emphasise that it is the nervous systems, and not the epiphenomena which accompany nervous action, that help to produce behaviour. I conclude that Popper's attack on epiphenomenalism appears to rest on the tacit assumptions that behaviour (which influences culture and the physical world) is influenced by conscious states, and that the evolution of behaviour is due to the evolution of conscious states. I have argued that these assumptions can be dispensed with.

Notes

1 K.R. Popper, 'Natural Selection and the Emergence of Mind', The First Darwin Lecture given at Darwin College, Cambridge, 8 November 1988.

2 G.D. Wassermann, 'Testability of the Role of Natural Selection within Theories of Population Genetics and Evolution', The *British Journal for the Philosophy of Science* **29** (1978), 223-242.

3 K.R. Popper and J.C. Eccles, *The Self and its Brain* (Berlin: Springer International, 1977).

4 T.H. Huxley, *Method and Results: Collected Essays Volume 1* (London: Macmillan, 1898) p. 240.

5. Cf. G.D. Wassermann, *Molecular Control of Cell Differentiation and Morphogenesis* (New York: Marcel Dekker, 1972); G.D. Wassermann, 'Molecular Genetics and Developmental Biology', *Nature New Biology*, 245, (1973), 163-165 for earlier models and op. cit. in note 6, below for a much improved theory of the development of nervous systems.

6 G.D. Wassermann, *Neurobiological Theory of Psychological Phenomena* (London: Macmillan, 1978).

7 See Sec. 4.7 in G.D. Wassermann, *Brains and Reasoning: Brain Science as a Basis of Pure and Applied Philosophy* (London: Macmillan, 1974).

University of Newcastle Upon Tyne

7 Thoughts on Popper

Halstead[1] has mentioned Popper's recantation[2] of some of his mistaken ideas on evolution. The unmasking of some of Popper's apparent errors occurred as follows. In May 1977 my paper[3] on 'The testability of the role of natural selection within theories of population genetics and evolution' was widely circulated and received for publication. In that paper I explained at length why Popper's ideas on the testability of natural selection seem totally wrong. Popper's own recantation appeared first somewhat later in his Darwin lecture (Darwin College, Cambridge, November 1977). His brief arguments seemed to be as wrong as ever and were based on isolated instances of natural selection, instead of referring to some population-genetic theories that involve natural selection. The Darwin lecture also contained an apparent flaw as regards Popper's views on epiphenomenalism[4] (these views were based on evolutionary considerations). Ruse had earlier criticized Popper's apparent misconceptions on evolution[5], but further scrutiny was called for. Hence, in a recent paper[6], following the earlier critiques by Lewontin and Ruse of Popper's views on evolution, I analysed at great length the likely nature of the 'theory of evolution'. In that paper I noted that 'it would be wrong to think that Popper has correctly assessed what matters in evolution', and I tried to exhibit in great detail the apparent fundamental errors of many of Popper's ideas on evolution, although my paper is mainly constructive.

Notes

1 L.B. Halstead, *Nature* **292**, 403-404 (1981).
2 K. Popper, *Dialectica* **32**, 344 (1978).
3 G.D. Wassermann, *Br. J. Phil. Sci*, **29**, 223-242 (1978).

4 G.D. Wassermann, *Mind* **88**, 572-575 (1979).
5 M. Ruse, *Phil.Sci* **44**, 638-661 (1977).
6 G.D. Wassermann, *Phil.Sci* 'On the Nature of the Theory of Evolution' ***48***, 416 - 437 (1981).

University of Newcastle Upon Tyne

Addendum

A copy of reference six can be found in *KEYS TO LIFE, Philosophy and New Mechanisms of Evolution and Development*, Ashgate Publishing, Aldershot 1997.

8 A critique of some aspects of human ethology

Eibl-Eibesfeldt (1979) (=E-E) regards sociobiology as 'a basic constituent of ethology'. I shall therefore examine some of the common weaknesses of current human ethology and human sociobiology. Ethological studies of animals often deal with species-specific, largely stereotyped, forms of behaviour - say, fixed-action patterns in mating displays - which are not encumbered by superimposed, fickle, cognitive activities. The species-specific stereotypy of such behaviour makes us believe that it is essentially genetically determined, even if the genes remain unknown. On the other hand, human behavioural and cognitive capacities, that can be used for very fickle types of behaviour, are also likely to depend on large multitudes of genes, whose combinatorial actions could help to control the development of nervous systems, thereby providing the machinery for these capacities (cf. Wassermann 1978a, chap. 2). Thus all human behaviour and cognitive capacities depend on genes, because they depend on nervous systems and their development. This leads to serious issues. Are human ethologists and human sociobiologists mainly interested in evolutionary and descriptive aspects of the kind of behaviour that supposedly can be recognized as strongly 'heritable' without probing into the nature of the genes involved? If so, the absence of an assured genetic basis might lead to faulty conclusions vis-à-vis those who believe that most or all human behaviour is learned. On the other hand, if human ethologists and human sociobiologists wish to found their sciences on precise genetic knowledge, they might be confronted with monstrous or unattainable goals. It is some of these issues that I wish to discuss.

Although animal sociobiology has progressed with various studies of group selection, kin selection, and so forth (cf. Maynard Smith 1976; Wilson 1975; Parker 1978), the value of insights and the scope of applicability of human sociobiology as propounded by Wilson (1975, 1978) and others have been called into question (Maynard Smith 1978; Rose 1978a, 1978b; Parker 1978). Human aggressiveness, altruism, intelligence, social competitiveness, territorial group-living habits, male dominance features, among others, are supposedly 'partly heritable', but the degree of 'heritability' is either unknown or is based, where possible, on *ad hoc* measures such as IQ test comparisons in the case of intelligence. The notions of 'heritability' and 'partial heritability' unless backed up (as in observations of animal ethology) by independent criteria, are apt to remain vacuous in the absence of a firm genetic basis. This was aptly recognized by Feldman and Lewontin (1975), who severely criticized arguments supposedly favouring the alleged 'heritability' of intelligence - the latter being based on IQ tests (Jensen 1972, 1973). Cognitive psychologists cannot claim that they have hitherto adequately characterized the meaning of intelligence. Feldman and Lewontin (1975) stressed that 'there is a vast loss of information in going from a complete machine (such as the human brain) to a few descriptive parameters (such as IQ and its 'heritability'). Therefore there is immense indeterminacy in trying to infer the structure (and capacities) of the machine from those few variables, themselves subject to error.' (Feldman and Lewontin's remarks, however, must not lead us to conclude that studies of experimental psychology could not provide powerful leads for neuropsychological theories, even if, like all theories, they remain very fallible).

More generally, behavioural science measures that are based on a few statistical parameters fail, as a rule, to assess intricate cognitive and other complex human capacities, and they can tell us little, if anything, about the manner in which these capacities can be utilized, particularly in motivationally obscure situations If there should exist preferences for wine, women, and song that are supposedly in part genetically encoded, then we must either find the relevant genes or other unambiguous indicators of genetic encoding. This alone could elucidate what E-E (p. 7) terms the 'prospective potential'. When E-E claims (p.8) that 'man is certainly programmed to act in predictable ways, and much of his behavioural repertoire is clearly acquired during ontogeny', does he wish to apply this statement to complex cognitive processes? If so, where is the evidence? I suspect that E-E is unjustifiably replacing genetically controlled, developmentally established, capacities of man to acquire behavioural repertoires (by learning) by the putative developmental acquisition of these repertoires themselves. E-E's references to hypothetical human innate releasing mechanisms and other supposedly inherited human behavioural tendencies refer to nearly stereotyped behaviour (cf. E-E, p. 9) in which developmental programming might have occurred. One cannot, however, assume that complex human cognitive behaviour is remotely akin to such stereotyped behaviour. Hence, nearly stereotyped forms of human behaviour could have few, if any, implications for the sciences of man, which depend strongly on intricate human cognitive capacities.

I suggested already that most human behaviour-controlling genes might code principally for the development of nervous systems, and thus for capacities rather than actual behaviour programmes. Behaviour-controlling genes could also be expected to code for the development of auxiliary structures and organs used in behaviour. (See Wassermann, 1972, 1973, 1978a, Clowes and Wassermann 1984 for theories concerning the ways in which genes could control the development of nervous systems, and could thereby account for many findings reviewed by Jacobson 1970, Gaze 1970; Hunt and Jacobson 1974; Wassermann 1978a). Yet gene control of the development of human nervous systems does not necessarily determine in detail what future use people can make of their nervous systems. That man has the (presumably innate) capacity to learn does not tell us what people will learn. Human capacities for exhibiting hunger, thirst, sex urge, and so forth are, you can be sure, partly genetically controlled. Yet, what dishes a hungry person orders in a restaurant may largely depend on learning or other contingencies, although partially genetically determined food preferences simply cannot be ruled out. Instead of being able to rely on known genetic systems and their actions, E-E, in common with other human ethologists, depends, in the absence of other criteria, unenviably, on cross-cultural comparison, on extrapolations from studies of infant behaviour, and on analogies with animal studies none of which seem to be remotely as reliable as the procedures of animal ethology.

E-E (p. 2) claims that human ethology is based upon the theory of selection and inquiries, for example, as to what selective pressures have operated to bring an observed structure into being, on the assumption that the structure somehow contributes to the inclusive fitness of its possessors. Such a logically 'existentially quantified' statement rests on untestable assumptions concerning the hypothetical structural contribution to inclusive fitness. Moreover, we do not know how many kinds of human genes and their alleles are involved in human behaviour, and how many of these alleles could be selective neutral (see Wassermann 1978b for a brief discussion of the selectively neutrality hypothesis). For all we know, selectively neutral alleles of many genes might contribute to the control of the development and the functions of human (and animal) nervous systems. It would therefore be futile to ask what selective pressures have acted on these alleles in the gene pool (via phenotypes). E-E's evolutionary arguments are also debatable in other respects. His claim (E-E, p. 8), that 'learning is determined by phylogenetic adaptations, such that animals learn what contributes to their survival and adaptively change the behaviour through experience', suggests that everything that animals and people learn has survival value, which seems to be an irrefutable hypothesis. People could always argue on an *adhoc* basis that any particular item that they have learned increases their fitness, and how could one refute that this is the case? Significantly, Parker (1978) has noted that 'a major weakness of current sociobiological theory is the omission of insights into the evolutionary dynamics of animal learning...there is little or no formal modelling of the learning process...The act of reinforcement itself must be an evolved phenomenon...'

Surprisingly, E-E (p. 22) suggests that 'there exists a universal "grammar" of human social behaviour whose study will open promising avenues of co-operation between ethologists and sociolinguists in the field that might appropriately be dubbed "etholinguistics".' This claim goes even far beyond Chomsky's (1972) controversial views concerning possible innate aspects of deep structures of grammar. Nevertheless, some findings of human ethologists seem to be of genuine, but possibly exaggerated, scientific interest. It would therefore be unfortunate if E-E's numerous, partly metaphysical and partly unjustifiable claims were to tempt many scientists to dismiss his subject as worthless. (See also Chomsky, *BBS* 3(1), 1980).

From *Behavioral and Brain Sciences* (1980) **3**, 630-631.

Bibliography

Bullock, T.H. (1970) *The reliability of neurons*, Jacques Loeb Memorial Lectures, Woods Hole Oceanographic Institution and Marine Biological Laboratory, Massachusetts

Chomsky, N. (1972) *Language and mind*, New York, Harcourt Brace Jovanovich

Chomsky, N. (1980) *Behavioral and Brain Sciences* **3**, 1

Clowes, J.S. and Wassermann, G.D. (1984) Genetic control theory of developmental events, *Bulletin of Mathematical Biology*, **46**, 785-825

Eibl-Eibesfeldt, I. (1979) 'Human ethology: Concepts and implication for the sciences of man', *Behavioural and Brain Sciences*, **2**, 1-57

Feldman, M.W. and Lewontin, R.C. (1975), The heritability hang-up, *Science* **190**, 1163-68

Gaze, R.M. (1970) *The formation of nerve connections*, London, Academic Press

Hebb, D.O. (1949) *The organization of behavior*, New York, John Wiley

Hunt, R.K. and Jacobson, M. (1974) Neuronal specificity revisited, *Current topics in developmental biology*, **8**, 203

Jacobson, M. (1969) Development of specific neuronal connections, *Science* **163**, 543-547

Jacobson, M. (1970) *Developmental neurobiology*, New York, Holt Rinehart

Jensen, A.R. (1972) *Genetics and education*, London, Methuen

Jensen, A.R. (1973) *Educability and group differences*, New York, Harper and Row

Maynard Smith, J. (1976) 'Group selection', *Quarterly Review of Biology*, **51**, 20-30

Maynard Smith J (1978) 'Constraints on human behaviour', *Nature*, **278**, 120-21

Palay, S.L. (1967) Principles of cellular organization, in *The Neurosciences*, eds. G.D. Quarton, T. Melnechuk and F.O. Schmitt, New York, Rockefeller University Press

Parker, G.A. (1978) 'Selfish genes, evolutionary games and the adaptiveness of behaviour', *Nature*, **274**, 849-855

Perkel, D.H. (1970) Spike trains as carriers of information, in: *The Neurosciences second study program* ed. F.O. Schmitt, New York, Rockefeller University Press

Rose, S. (1978a) Essay review of E.O. Wilson's book *On human nature*, *Race and Class* **20**, 277-287

Rose, S. (1978b) Review of E.O. Wilson's *On human nature*, *New Scientist* 19th October

Sommerhoff, G. (1950) *Analytical biology* Oxford, Oxford University Press

Sommerhoff, G. (1974) *Logic of the living brain*, New York, John Wiley

Wassermann G.D. (1972) *Molecular control of cell differentiation and morphogenesis* New York, Marcel Dekker

Wassermann G.D. (1973) 'Molecular genetics and developmental biology', *Nature New Biology*, **245**, 163-165

Wassermann, G.D. (1978a) *Neurobiological theory of psychological phenomena*, London, Macmillan

Wassermann, G.D. (1978b) 'Testability of the role of natural selection within theories of population genetics and eveolution', *British Journal for the Philosophy of Science*, **29**, 223-242

Wassermann, G.D. (1979) Reply to Popper's attack on epiphenomenalism, *Mind* **88**, 572-575

Wassermann, G.D. (1981a) Book review of *Teleology* by Andrew Woodfield, *Philosophia* **10**, 125-132

Wassermann, G.D. (1981b) On the nature of the theory of evolution, *Philosophy of Science* **48**, 609-628

Wassermann, G.D. (1982a) Materialism and mentality, *Review of Metaphysics* **35**, 715-729

Wassermann, G.D. (1982b) TIMA part 1. TIMA as a paradigm for the evolution of molecular complementarities and macromolecules, *Journal of Theoretical Biology* **96**, 77-86
Wassermann, G.D. (1982c) TIMA part 2. TIMA-based instructive evolution of macromolecules and organs and structures, *Journal of Theoretical Biology* **99**, 609-628[1]
Wassermann, G.D. (1983a) Human behavior and biology, *Dialectica* **37**, 169-184
Wassermann, G.D. (1983b) Quantum mechanics and consciousness, *Nature and System* **5**, 3-16
Wilson, E.O. (1975) *Sociobiology: the new synthesis*, Cambridge Mass., Harvard University Press.
Wilson, E.O. (1978) *On human nature*, Cambridge Mass., Harvard University Press

Note

1 See also Wasserman, G.D. (1997) *Keys to Life*, Aldershot, Avebury, Ashgate Publishers Ltd., Chapter 6 for TIMA part 3, without which the TIMA theory is incomplete.

9 Book review of 'The Chief Abstractions of Biology' by W.M. Elsasser (North Holland and American Elsevier)

This review was done at the invitation of Professor Lord Walton (now Oxford University) for the *Journal of the Neurological Sciences* **30**, p. 216 (1976), a journal of which Lord Walton was then editor.

This highly self-repetitive book provides no new insights into biological phenomena, but tries to disseminate antimechanistic bio-polemics, leaning sometimes on the ideology of other eminent contemporaries. Elsasser's choice of 'chief abstractions' of biology (*viz.* the vague concepts: 'contingencies', 'variability', 'complexity', 'heterogeneity' and 'individuality') is markedly idiosyncratic. His abstractions concern the hard core of almost or completely unsystematizable variations inherent in any samples drawn from a class of living or inanimate entities. He makes theoretical mountains out of circumstances which arise usually from experimental mole hills, and his abstractions are the antithesis of the precise conceptual frameworks of molecular biologists and other mechanists in the life sciences. Elsasser repeatedly assures us that he is no expert on various topics which he discusses. This assurance seems to be vindicated by his tendency to rely on dogmatic apriorisms and by several serious misstatements.

For example, Elsasser claims (p. 122) that

> there is extensive evidence to the effect that... [a mechanism whereby hereditary transmission of morphological properties is effected] does not exist.

The 'extensive evidence' is not provided, and it is hard to see what positive experimental evidence could be obtained for the non-existence of an unknown mechanism. On the contrary, Garcia-Bellido *et al.*, (1973, *Nature New Biology*, **245**, 251) and P.A. Lawrence have demonstrated that specific homeotic genes control particular morphological mechanisms in Drosophila imaginal disc development.

Organismic variability and contingencies, on which Elsasser concentrates, are familiar to mechanists, who expect them and deal with variability statistically where possible. They realize that similar biological mechanisms, notably biochemical ones, may utilize genetically variant components in different organisms of the same species. These variant components may still be functional, but may operate suboptimally. Faced with this variability, mechanists try to see the wood for the trees. Their experiments and idealizing models concentrate on essentials of mechanisms and attempt to *minimize* the influence of genetic and environmental contingencies, where this is possible or desirable. By using a variety of experimental techniques, mechanists can make reasonably sure that the contingencies due to experimental artefacts are gradually being eliminated in many or most cases. The residue of contingencies appears then to play a negligible part, so that it is of purely academic interest to note Elsasser's truism that mechanisms cannot be (completely) disentangled from contingencies (p. 28). His advice that 'indeterminacies', related to contingencies, should, if possible, be *maximized* (p. 20) by use of inductive probabilities may seem unrealistic to most experimentalists.

Individual variability of organisms, which Elsasser regards as crucial, has long been known to be partly a consequence of the presence of numerous alleles for genes in the population. Gene mutations and nutritional (etc.) variations are also familiar contingencies. In fact, alleles have thrown much light on certain genetic mechanisms! Yet, some molecular biological experiments are designed so that allelic effects can be disregarded (cf. Galau *et al*, 1974, *Cell* **2,** 9).

The persistence of many mechanisms throughout evolution, thanks to natural selection, is more significant than their variability. In particular species-specific developmental mechanisms are of great prominence, since they help to ensure the reappearance of individuals of each species, and with it species-specific behaviour. Elsasser would hardly deny that, despite allelic variations and other contingencies, developmental programs are being translated into individuals endowed with species-specific features. For instance, despite contingencies, developmental mechanisms can build up nervous systems which are species-specifically wired-in. This can provide behaviour mechanisms of exquisite species-specificity, whose effects are being widely studied by ethologists. Thus, K. Lorenz (1965, *Evolution and Modification of Behaviour*, p. 25) cites that:

> A young male salticid spider which, after moulting for the last time, approaches a female, must neither mistake another species for his own nor must he perform the signalling of his courtship dance in any other way than the one to which his female responds; otherwise he would be eaten immediately. He has no opportunity in his short life to gain any information

about what a female of his species looks like, nor what movements he must perform to inhibit her feeding reactions and to stimulate her specific mating responses.

Life scientists would like to explain the underlying intricate mechanisms of this behaviour. Knowledge that no two salticid spiders are exactly alike is useless in resolving such mechanistic problems. Elsasser's abstractions lack explanatory value, and obscure the essentials of mechanisms. Contrary to his intention, his book may help to convince people that the 'organicism' and 'holism' which he advocates are relics of the past, whereas the future lies with molecular biology and other mechanistic investigations as well as with mechanistic theories.

10 Kindly explained

Marks[1] as well as his creationist critics[2-5] seem to be arguing at cross purposes. Unlike Marks, one ought not to sneer at the views of fundamentalists, just as a clinician must not disregard the delusions of a psychotic. Both types of belief mean much to those who hold them. A creationist cannot be convinced by facts or scientific arguments of any possible errors of an *a priori* infallible creed, any more than a psychotic can be shaken about the absolute truth of his delusions by rational counter-arguments. When contrasting science and creationism we must remember that we are dealing with different *kinds* of explanations concerning the world. Let me paraphrase in an evolutionary context a passage from the late C.A. Mace written in a behaviour-scientific context in an editorial foreword to a book by Vernon[6] .

> The difference between evolutionists on the one hand and creationists on the other arises perhaps from a failure of the latter to do justice to the fact that there are many different *kinds* of explanations. There *is* the kind of explanation that fossils exist, which simply says that fossils are remains of past organisms, irrespective of whether these organisms were divinely created or arose by self-organization. This is a neutral kind of explanation. There *is* the sort of explanation which appears to satisfy evolutionists and some other life scientists, which takes the form of saying that fossils are relevant to our formulations of particular theories of evolution-specific mechanisms[7]. There is another kind of explanation which satisfies materi-alistically-minded mechanists, who (like myself[8]) suggest that fossils are the remains of organisms that arose by self-organization. Perhaps also some notice should be taken of the kind of explanation which satisfies children and devout old ladies who would say that fossils exist because God arranged

things so as to remind us of the many beautifully adapted organisms that he has created in the past.

There is room for all of these views among the right people in the right places and in mutually exclusive ways. We must remember that different kinds of explanations rely on different premises and on different explanatory procedures. Thus, hypothetico-deductive scientific theories are not intended to proclaim absolute truths, but are, or should be, falsifiable. By contrast, the doctrines of creationists are proclaimed as absolute truths of a metaphysical kind, and do not function like scientific hypotheses. If scientific facts do not fit creationist doctrine then, with sufficient ingenuity, an indefinite number of reinterpretations of the doctrinal metaphysical statements remain possible. In the Popperian sense there exists an unlimited range of built-in metaphysical defences of fundamentalist doctrines. It is therefore not uncommon for scientists to accept two schemes which may be related to the same facts. On the one hand they may accept hypothetico-deductive theories of mechanisms. On the other hand they may embrace a metaphysical belief system which explains some of the facts known to them from science in ways totally different from the ways in which they occur within scientific theories. I know many such people and some highly intelligent ones among them, although I see no personal need to supplement science with a *fundamentalist* metaphysics. (Some kind of metaphysics is probably needed in science[9].)

Notes

1 J. Marks, *Nature* **295**, 276 (1982).
2 F.W. Cousins, *Nature* **295**, 452 (1982).
3 N.K. McBride and L. Kreitzer, *Nature* **296**, 8 (1982).
4 A.J. Hollin, *Nature* **295**, 548 (1982).
5 T.H. Hamblin, *Nature* **296**, 108 (1982).
6 M.D. Vernon, *The Psychology of Perception* (Penguin, London, 1962).
7 G.D. Wassermann, *Phil.Sci* **48**, 416-437 (1981).
8 G.D. Wassermann, *J.Theor.Biol.* **96**, 77-86 (1982).
9 G.D. Wassermann, *Review of Metaphysics* **35**, 715-729 (1982).

11 Book review of *The biological origins of human values* by George Edgin Pugh (Routledge and Kegan Paul, 1978)

Few people, I suspect, will have the patience to read from cover to cover this dilettantish, oversimplified and overextended excursion into the already hotly debated fields of human sociobiology and human ethology[1]. Pugh's leitmotif arises from the hypothesis that human actions are partially instigated or controlled by value-driven decision systems which rely on 'innate' values. In addition, acquired 'secondary' values are also assumed to guide our behaviour, thereby implying, that values that are not innate must be acquired, which is not a startling revelation. Pugh's contributions lean predominantly on vague speculations, which often lack adequate empirical backing, and which provide, at least in my opinion, no novel insights or deductively fruitful explanatory constructs. I found much of the text to be redundant and to consist of long sequences of *ad hoc* assumptions and should be surprised if it offers anything attractive to sophisticated philosophers. To scientists it presents mainly untestable hypotheses. Indeed, it frequently simply reclassifies well known phenomena within a superficial decision-theoretic framework. Many things which seem 'clear' or 'obvious' to Pugh, might appear far from clear or obvious to sceptics. The style and level of arguments of the book seem to me too elementary for experts, and too vague and misleading for laymen who would form the most appropriate readership. I think that a short paper of 30 printed pages could have conveyed the essentials of Pugh's theses without having to pay for 450 pages which, in my opinion, contain many platitudes as well as various errors. Let me substantiate these remarks by selected examples.

Pugh's 'theory' of values 'had its origin in cybernetics and decision science' (p. IX). He aims to show that the concept of a value-driven decision system 'provides a unifying framework that links both behavioral science and human values to the evolutionary design of the brain as a biological control system.' Allegedly this 'theory helps to explain how the genetically inherited behavioural tendencies identified by sociobiologists can operate through the rational mind to motivate conscious human behaviour.' (p. XI). Those who regard *human* sociobiology as a pseudo-science might therefore discard the book without reading it. But as a reviewer I persisted from cover to cover. From the start Pugh places strong emphasis upon genetic control of behaviour (at least by implication), without ever telling us how genes could indirectly control the machinery of behaviour. He claims (p. 6) that 'When the brain is analyzed as a value-driven decision system, then our fundamental "human values" seem to follow naturally as an almost inevitable consequence of the "design-requirements" for such a value-driven system. They are an essential part of nature's system design for the brain.' This level of discourse, which is typical of practically every part of the book, is unlikely to cut ice with experimental neuropsychologists, who are in search of detailed *mechanisms* rather than simplifying platitudes. When Pugh (p. 7) claims that 'there is strong evidence that the evolution of the brain has been guided by the same basic principles of efficiency that dictated the value-driven design concept for the artificial decision systems' then he seems to confuse what has evolved with *how* this evolved, since we know little, if anything, reliable about the selective pressures involved and the molecular biological events of brain evolution.

Pugh distinguishes primary 'values' (e.g. sensations of unpleasantness, pain and hunger) which are built in (p. 8) from secondary values (e.g. moral values) which are acquired. He believes that 'scientific methods (based on fundamental knowledge of human decision systems) can assist in deciding (in terms of fundamental human values) what is desirable or what is right.' This seems to me quite wrong. With luck science might discover *how* human values are represented by brains, but science cannot tell us which values we ought to adopt (e.g. which moral values). Science is not ethically normative. We are told that 'The "decision-system" concept interprets human behaviour as a purposive or goal-seeking activity' (p. 19). But this just shifts the problem of explanation, since one has to explain what is meant by 'purposive' or 'goal-seeking', i.e. it raises all the contested issues of teleology. To argue (p. 20) that 'it seems quite natural to think of the human brain as a "decision system"' does not explain how the human brain represents decision processes. Apart from this, brains help to perceive, to form concepts, to learn, and many of these activities do not depend on deliberate decisions, but are automatic (e.g. perception, visual illusions etc.).

Pugh's approach to evolution is unsophisticated, particularly when he discusses natural selection. Then there are numerous straight errors. Thus (p. 24) 'Both inorganic chemistry and organic chemistry have been "explained" by quantum mechanics'. But quantum mechanics, even when aided by substantial empirical knowledge, has hitherto not explained the details of even the 'ground state' of any major protein, say, haemoglobin. On p. 28 Pugh describes how artificial decision systems operate by assigning numerical 'values' to each alternative. This sets the tone for much of the rest of the book, where he frequently interprets the brain (or at least part of it) as if it were a computer. This simple ideology has long been abandoned by most neuropsychologists. The fact that artefacts can achieve a representation *X* by computing does not mean that brains achieve a representation of something denoted as '*X*' by 'computing'. In fact, Pugh never specifies how brains act as alleged computers. For him the word 'computer' serves as a kind of magic wand. How, for instance, does he know that brains 'compute' values, i.e. that they represent 'values' in terms of machine states which represent numbers? For Pugh the word 'value' refers 'simply to a quantitative criterion of decision' (p. 38). Yet, in an artificial decision system the designer must make prior decisions by allocating numerical values to alternatives, i.e. he already relies on his own brain evaluations. Hence, to take the artificial system as a model for a brain decision system could be circular, unless the brain's numerical valuation machinery is explicitly described, which is not the case. *Carte blanche* is obtained by asserting (p. 30) that 'innate values are not constant over time, but they respond to physical and social stimulation in accordance with complex built-in rules'. These rules are not stated. We are told (p. 30) that 'the instinctive values include a mix of both altruistic and selfish elements'. Such vague all-explanatory statements could be expected of a weak text on human sociobiology.

In chapter 2 Pugh reaches the unremarkable conclusion that brains are not randomly wired-in. But he does not discuss the complex issue of the degree of precision of the wiring-in in relation to perturbing non-genetic factors. When Pugh (p. 42) writes about nervous codes he restricts himself implicitly to spike-train codes, leaving aside other possibilities[2]. The main problems of coding seem to have eluded his attention. He also has taken no notice of the serious criticisms raised by Moray[3] and others against the interpretations of the experiments of Hernandez-Peon and his associates to which Pugh (p. 46) refers. Again, Pugh (p. 47) suggests that we can deduce from Penfield's brain stimulation experiments 'that the memory traces may indeed be maintained...in the most compact and selective form'. Since nobody knows or could know for sure what could correspond in nature to the most compact and selective form of a memory trace, it is hard to see how Pugh's conclusions could follow from any observations.

On p. 59 we are told that 'The assignment of *primary* values [in man]...occurs *outside* the veil of consciousness.' But on p. 60 we read that 'The primary human value system is experienced by the individual in the form of specific valuative sensations of physical and psychological origin'. What are we to make of such apparent stylistic contradictions? The very frequent invocation of evolution as the generator of decision systems makes the reader aware of a sizeable evidential gap between Pugh's claims regarding artificial decision systems and his claims concerning biologically evolved decision systems. Analogy seems to be essential to many of Pugh's arguments. There occurs also a tendency to interpret 'drives' as values, and at other times to consider sensations associated with drives as 'values' (cf. p. 112). Then there are vacuous statements like (p. 117) 'Presumably the signal [which is the physical correlate of an innate value] corresponds to a specific pattern of neural discharges in a specific set of neurons (which could in principle be measurable).' Equally vacuous is the existentially quantified statement (p. 118) that 'Each of the innate human value sensations is generated by some biological mechanism within the human brain.' Such statements are untestable. In fact, pp. 118-119 contain a gamut of similarly vague statements. On p. 120 we are informed that 'very efficient neurological mechanisms exist for data storage which can permit efficient computer-like learning in central control systems.' How does Pugh know that neural memory systems are computer-like? In fact, Pugh's neurobiological writing seems often misinformed. He disregards axo-somatic synapses (p. 126). Again he tells us (p. 127) that 'the basic functional design is very similar for all neurons.' What about those myriads of vertebrate visual receptor neurons which do not generate nerve impulses when activated? Then there are absurd sentences (p. 127) 'The resulting information (a signal or non-signal) is transmitted via the axon.' How can a non-signal be *transmitted* via an axon? Or how do we know (p. 127) that 'Data-processing logic occurs in the dendrite (input) zone which determines when neurons will fire?' Surely the synaptic input to the soma is also instrumental. On p. 128 we are told that 'It seems clear that with a little practice computer designers could learn to use neurons as circuit elements in place of transistors, assuming that methods could be found for hooking them up in accordance with desired specification.' There exists no evidence to support the wild speculation that neurons act like transistors.

On p. 131 we are told, on an *ad hoc* basis, that 'the hypothalamus operates as a special purpose computer which provides values for the decision system', and that the 'value driven decision system...is located in the forebrain'. Worse still, Pugh confuses mind and brain (p. 131) when he tells us that 'The conscious mind...lies above the hypothalamus and includes almost all of the rest of the forebrain.' Again supposedly (p. 132) 'The sensation of consciousness...seems to arise within the decision systems', which Pugh (p. 131) allocates to the forebrain. But surely we are also conscious of colours, and are aware of shapes, etc. Are these allegedly represented in the forebrain? I find Pugh's speculations reminiscent of phrenology of the worst kind. On p. 133 we are told that 'In the human body, the processing required to transform the action decisions of the mind into detailed muscle commands normally occurs outside the "mind". It is accomplished primarily in an output processor called the cerebellum...' Does Pugh not accept that the processing of human speech is at least partly related to events in the cerebral cortex of the left hemisphere? When Pugh (p. 134) writes 'Evidently, the forebrain houses a 'rational' decision system...' this may be evident to him, but what is the evidence? The word 'orchestration' occurs frequently in the book. Thus, 'The control of orchestration of the value system is centred in a special computation area in the vicinity of the hypothalamus' (p. 137). Elsewhere Pugh (p. 138) writes 'Once the signals have reached this area of the cortex they are available to be interrogated by the conscious mind.' Does he wish to advocate neo-Cartesian interaction dualism? In another context (p. 145) Pugh claims that 'the mind operates like an "associative processor".' What does this mean? And what evidence is there that 'The dominant verbal hemisphere specializes in symbolic logic...and [that] its association processes are essentially digital', whereas 'the other hemisphere may operate more like an analogue computer' (p. 149)? Pugh's major escape clause appears on p. 153 where he apologetically states 'We have not dwelled on our vast areas of ignorance about the brain. Consequently there is a risk that the reader may get the false impression that the human brain is well understood...' If so, why spend so much space on numerous *ad hoc* speculations and absurdities?

Chapter 7 is devoted to consciousness and free will among others. Pugh does not discuss any form of the identity theory, but believes (p. 154) that 'it seems likely that consciousness itself may have evolved as a natural (perhaps inevitable) by-product of the value-driven decision system.' This is pure speculative metaphysics. Again, he claims (p. 157) that 'The interpretation of the human brain as an automatic decision system provides a rather simple resolution (at least in scientific terms) of the long-standing controversy concerning "free will" versus "determinism" '. He refuses to define "free will" (p. 158), but argues that a decision system makes choices. Yet this simply implies the tautology that a decision system decides. Pugh ties himself into knots when it comes to "free choice" (pp. 158-159), and opts for free choices being made relative to a primary decision criterion, based on an *a priori* value system. But this will not do, since his value system is postulated, and has not been empirically isolated, so that Pugh's reduction of "free choice" is of the explanation *ignotum per ignotius* variety. Again he claims (p. 159) that 'Although we did not choose our innate values we certainly do make choices in terms of these values. The choices are as "free" as is theoretically possible for a "choice" to be'. But Pugh does not tell us what he means by 'theoretically possible', which could mean either completely 'free' or 'not free' at all, so that he evades the issue by his verbal game.

Elsewhere we are told (p. 173) that the 'innate' [human] values include both emotions and the traditional *biological drives*. The drive concept itself is questionable, and to equate drives with values is somewhat like equating apples with apple trees. There occur other extraordinary statements and claims. Let me just cite a few more. Thus (p. 178) 'Each species (and each individual within a species) seems to have a characteristic motivational profile, just as it has a characteristic facial profile.' What does Pugh mean by a 'motivational profile'? On pp. 187 ff Pugh suggests a supposedly quantifiable theory of innate drives of infant primates, but does not supply this theory in explicit detail. There occurs a typical platitude (p. 192) when Pugh claims that 'The infant decision system has at its command the neural control levers necessary to initiate body movements, but the infant decision system does not know how to use the controls'. Again, consider the statement (p. 194) 'Fear in the infant begins as an innate response to specific genetically inherited stimuli'. Does Pugh think that stimuli are inherited? On pp. 218-219 Pugh defines 'value functions' for specific values in an *ad hoc* manner. His claim that human incest taboos are partly innate (p. 233) is questionable since there exists no reliable relevant evidence. To assert that a bee gets 'angry' (p. 271) is pure anthropomorphic writing. Also, *pace* Pugh (p. 278), Lorenz, Tinbergen and von Frisch have not demonstrated the evolutionary basis of behaviour, but the 'innate' (i.e. partly genetically conditioned) nature of *some* behaviour.

On p. 300 Pugh asserts that after wearing goggles which produce differential spectral refraction, and after dispensing with these goggles there occur aftereffects because, allegedly, 'the synapses in the optic nerve automatically adjust their interconnections so that *different* colours in the visual image are brought into the best possible registration'. There exists no unambiguous experimental evidence for this statement. Again, the fact that a stimulus-enriched environment may lead to an increase in cortical weight (p. 302) need not imply formation of new synapses, but might only involve augmented intraneural (and inraglia) protein and RNA syntheses (as has been found) and, hence, increased size and weight of neurons and glia concerned. Also, *pace* Pugh (p. 302), there is no evidence that the human skull enlarges during maturation as a *consequence* of a brain increase in size. Skull increase could be a consequence of an autonomous developmental process which continues into adult life. On p. 324 of the book there occurs an elementary error. Pugh states that there are almost 10 billion genes per individual. There are only ca. 3×10^9 DNA nucleotide pairs per human haploid set of chromosomes. How could these account for 10^{10} genes, with a gene requiring, in many cases, say, between 1000-1500 nucleotide pairs?

I could substantially augment this list of plain errors, *ad hoc* speculations and faulty philosophizing. But enough is enough, except for one final note. In part III of the book Pugh (p. 342) asserts that 'a mature psychology and sociobiology ought to provide practical guidance with regard to personal and social decisions'.

Yet, psychology and sociobiology are *sciences* (or partly pseudosciences) and unlike moral philosophy they are not concerned with individual or social behavioural norms. To discover and explain how individual and societies behave is not the same as trying to prescribe how individuals or societies ought to behave in specific circumstances, which is the task for judges, social workers, politicians and so forth. Philosophers who wish to get a glimpse of the poverty of Pugh's level of arguments concerning the applicability of his theory of human values to general social issues and to religion should turn to the section titled 'God's Will or Man's?' (p. 383).

Notes

1 J. Maynard Smith, 'Constraints on Human Behaviour', *Nature* **276**, 120-121 (1978); G.A. Parker, 'Selfish Genes, Evolutionary Games and the Adaptiveness of Behaviour, *Nature*, **274**, 849-855 (1978); I. Eible-Eibesfelt, 'Human Ethology - Concepts and Implications for the Sciences of Man' *Behavioral and Brain Sciences*, **2,** 1 - 57 (1979); E.O. Wilson, *On Human Nature*, Cambridge Massachusetts, Harvard University Press (1978); S. Rose, 'Review of E.O. Wilson's 'On Human Nature', *New Scientist* 19th October (1978). Recently the *Journal of Social and Biological Structures* (Studies in Human Sociobiology) has been put into production by Academic Press.

2 Cf. G.D. Wassermann, *Neurobiological Theory of Psychological Phenomena*. London, Macmillan (1978).

3 N. Moray, *Attention*. New York, Hutchinson Educational Press (1969).

12 Human behaviour and biology

Summary

Extremism in the environment-versus- innateness controversy in the behavioural sciences and in human sociobiology is being examined. Genetic effects can be severely modified or overruled by environmental factors, but may, nevertheless, be important. Dawkins' view that we are survival machines programmed to subserve selfish genes seems untenable and is a root of racialism. It is also argued that morality is compatible with mixed genetic and environmental control of brains via existing biological machinery.

Résumé

Est discuté l'extrémisme dans la controverse influence de l'environement contre innéité dans les sciences du comportement et la sociobiologie. Des effets d'origine génétique peuvent être considérablement modifiés ou supplantés par des facteurs environnementaux, mais ils peuvent néanmoins aussi être importants. L'idée de Dawkins selon laquelle nous sommes des machines à survivre programmées pour conserver des gènes égoïstes est intenable et source de racisme. L'auteur défend aussi la thèse que la moralité est compatible avec un contrôle mixte des cerveaux par des facteurs génétiques et environnementaux selon des mécanismes biologiques existants.

Zusammenfassung

Es werden die extremen Standpunkte in der Kontroverse, die sich in der Verhaltensforschung und in der menschlichen Sozialbiologie um die Frage Umgebung gegen Eingeborenheit dreht, untersucht. Genetische Effekte können durch Faktoren der Umgebung stark verändert oder verdrängt werden und trotzdem wichtig bleiben. Dawkins' Auffassung, wonach wir Überlebensmaschinen seien sollen, die für die Erhaltung egoistischer Gene programmiert sind, scheint unhaltbar und erweist sich als eine Wurzel des Rassismus. Es wird auch argumentiert, dass Moralität mit gemischter genetischer und umgebungsmässiger Kontrolle der Gehirne vermittelst biologischer Mechanismen vereinbar ist.

Among the frequently debated issues relating to the philosophy of human behaviour are the following prominent questions. First, how much of human behaviour is genetically and how much of it is environmentally conditioned? Second, in case that there should be demonstrable genetic components of behaviour then are complete behaviour modes in all their minutiae genetically (i.e. 'innately') determined, or does genetics only help to provide the basic machinery for executing the most diverse kinds of behaviour, leaving it to environmental contingencies to establish how the partly genetically provided machinery of behaviour is to be utilized? Third, to what extent do man's greatest assets, his intelligence (including power of decision-making etc.) and morality (where present), depend on his biological machinery or on his environment, and to the degree that they are biologically conditioned is the biological machinery comparable to that found for higher animal species? It is the aim of this paper to re-examine some of this often debated problematics.

Towards the end of a perspicuously argued paper that dealt with Dawkins' book 'The Selfish Gene'[1], Mackie[2] stressed that:

> We cannot simply apply to the human situation conclusions drawn from *biological models* [based on animal sociobiology]. Nevertheless they are significant and challenging as models; it will have to be shown how and whether human life diverges from them.

Independently of Mackie, I reached a similar conclusion, when arguing that findings in human ethology[3] were too trivial to base any far-reaching conclusions, extrapolated from animal ethology, on the available slender human studies. If the views of some animal sociobiologists, when extrapolated to humans, seem grossly exaggerated and gratuitous, and in some respects caricatures of Darwinism, this does not justify us to follow the emotion-tinted polemics of Mary Midgley[4] concerning Dawkins. As a moral philosopher she deserves much respect, but this does not mean that her biological arguments against Dawkins[5] which I shall not discuss here in detail are valid. In fact, any competent theoretical or experimental molecular biologist can readily rebut many of her technical arguments. Yet, she represents a school of thought that was very popular in certain circles in the pre-ethology days. For instance, she claims that 'within their friendship, mutual help (of animals) will indeed take place. But the reason will not be the recognition of an insurance premium falling due (as game-theoretical approaches of many population geneticists might suggest). It will be liking and affection'. To attribute specific feelings to animals, such as 'liking' and 'affection' is typical of many fairy tales, where animals can speak and are endowed with all kinds of human emotions and are supposed to have human values and human motivations (e.g. 'big bad wolf' planned to devour little red riding hood). This kind of talk, however, seems to me to be as gratuitous as the most excessive claims of certain sociobiologists about human behaviour, and I shall ignore it in the rest of this paper.

I

One of the central issues of ongoing philosophical, biological and psychological controversies is the old question 'how much of human behaviour is genetically, and how much of it is environmentally conditioned'? In answering such questions we have at one end, to cite Parker[6], 'the Skinnerian[7] view that almost all observable behaviour is the product of experience through reinforcement and conditioning.' At the other extreme are the opinions of those who believe that many phenotypic traits, notably behaviour traits, are predominantly genetically encoded and controlled. The latter kind of extremism has been advocated especially by people who believe that IQ has many genetic roots. It has been firmly rejected by others[8], but their arguments, in turn, are open to counterarguments[9]. For example, Thoday concluded that, when we disregard Burt's (apparently manufactured) data, we are 'still left with rather high heritability (of IQ), in the face of which the only arguments a determined environmentalist can use are to suppose that all the data are spurious, or to rest his case on the effects of environmental correlation between relatives mimicking completely the theoretical genetic correlations. Such environmental correlations must be a difficulty of human biometrical genetics, but it seems hard to believe that they could explain all the evidence, especially the evidence that the correlation between monozygotic twins reared apart (0, 74) is so much higher than dizygotic twins (0, 53) or sibs (0, 54) reared together... Such correlations, when assortive mating is allowed for, imply rather high heritability, which, however, is an upper limit, as some part of the correlation will be environmental'.

While the heritability controversy, concerning IQ, amounts really to one typical aspect of the question to what extent genotypes specify phenotypes, one may accept - with much caution - the assertion that IQ provides only a crude single parameter for defining measurable 'intelligence'. Yet, as Feldman and Lewontin[10] emphasized, 'there is a vast loss of information in going from a complete machine (such as the human brain) to a few descriptive parameters (such as IQ and its 'heritability') (see also[11]). Moreover, the relationship between genotypes and phenotypes is complex (see below), and it compelled Feldman and Lewontin to stress that:

> At present, no statistical methodology exists that will enable us to predict the range of phenotypic possibilities that are inherent in any genotype, nor can any technique of statistical estimation provide a convincing argument for a mechanism more complicated than one or two Mendelian loci with low penetrance. Certainly the simple estimate of heritability [proposed by some theorists], either in the broad or narrow sense but most especially in the broad sense, is nearly equivalent to no information at all for any serious problems of human genetics.

Pace Lewontin and Feldman's lucid argument, we must ask 'What can we really know about a complex machine, such as a human brain?' We can either talk at the level of 'typical' species-specific brains, which are idealizations for a wide class of brains of the species *Homo sapiens* or we can talk about what is typical about a specific brain that we have investigated partly either *in vivo* (e.g. by electro-stimulation[12]) or at the post mortem stage. When we talk about the class (or species) characteristics of a human brain, then we must always extrapolate, and consider an idealized phenotype. In this case we cannot really assert more than the typical or class characteristics of the performance capacities of such systems (i.e. brains). But how much use and what particular use any specific individual of the species can make of his brain depends on a host of contingent circumstances which are highly idiosyncratic, and, hence, unpredictable on the basis of known class characteristics. On the other hand it is by present methods not possible - nor in the foreseeable future - to investigate a particular brain with a view to finding out a great deal about the intelligence of its owner, even if we combined the brain investigation with psychological studies. This, perhaps, may be seen as the essence of Lewontin and Feldman's argument. They possibly wish to assert that in due course we may have access to many brain parameters in the *in vivo* brain, which at present are inaccessible, and which, combined with extensive psychological studies, might give us a better insight into the nature of human intelligence. For example, let us assume that combined psychological tests and - possibly post-mortem - biochemical tests revealed that human brains endowed with specific variants of a large set of structural proteins (or possibly some functional proteins) presented close correlates of intelligence. By sequencing the DNA of other kinds of human cells we might then be able to demonstrate that certain individuals in the population have the right combination of the correct gene alleles that code for a set of proteins that provide, as judged by psychological tests, for a certain (but not unambiguously exclusive) optimum of intelligence.

In this case we would have got much further than so-called IQ tests on their own. But this is very futuristic, although possibly not all that much.

As long as we stay at the level of behavioural generalities, we can already at present reach some awesome conclusions about relating molecular genetics and behavioural capacities of organisms. We may respectfully ask why a relatively 'lowly' organism, such as the sea urchin *Strongylocentrotus purpuratus*, requires some 20,000-30,000 structural genes[13] (of which only some 1000-1500 genes are ubiquitously required 'housekeeping genes' which code, presumably, for essential enzymes and other essential proteins) unless most of these genes code (combinatorially) for the structures of organs and support systems[14]. Moreover, recent evidence suggests that the great majority of proteins that are coded for by eukaryotic structural genes are '*structural proteins*' (i.e. proteins that help to organize structures such as cell membranes) whose genes have to be very accurately preserved in the population, since mutations at their genetic loci are highly deleterious, as witnessed by the sparsity of their alleles in the population[15]. This also suggests that there exists pretty stringent control - via a host of highly specific structural proteins - of the building plans of organisms. In particular, this applies to the wiring-in of nervous systems. Hence, one prerequisite for behaviour seems to be a genetically and, hence, biochemically pretty accurately specified, optimally obtainable, nervous system for each species.

Prima facie this conclusion seems to clash with the facts cited by Lenneberg[16] that 'human beings with dwarfed brains are often capable of perfect language'. Likewise there are Lorber's[17] observations. He writes:

> I have now studied more than 100 people who are among the most severe hydrocephalic cases, using the CAT scan (a computerized brain scanning technique), formal intelligence tests, and detailed neurological assessments. The majority show normal physical development as well as good intelligence. Among them are several with IQs of 130 or more who are doing splendidly at school. Some of the adults have university degrees, others are members of executive committees, one is a senior nurse, and most are in gainful occupations. They hear and see well. One young man has a first-class honours degree in mathematics at a university, despite the fact that some 95 percent of the cranial areas that normally accommodate the hemispheres are filled with fluid. Instead of the normal 4.5 centimeter thickness of brain tissue between the ventricles and the surface of the cerebral cortex, he has a layer measuring a millimeter or two.

How can one reconcile such facts with a supposed, normally well executed, structuring of an organism based on a genetically pretty nicely and finely controlled building plan? One possible answer (also suggested by Lashley's[18] extensive - although occasionally disputed - evidence concerning widespread equipotentiality of cortical and possibly also subcortical regions of various types of mammalian brains) is as follows[19]. Although cells may develop systematically into clones, according to a genetic programme, starting from the fertilized egg, clone formation may be disrupted environmentally. (A clone is a family of cells that descend all from the same ancestral cell, via a sequence of mitoses). In this case cells that, because of their genetic specification, normally seek out available optimal loci in relation to other cells (that recognize them because of their own specification, *via* genetically determined cell markers[20], will settle at suboptimally matching loci, when genetically specified optimally matching, and normally present, target cells are not available. In other words, the genetic programme now ensures a suboptimal alternative assembly programme of cells, which can still produce normally *functioning* (but not necessarily normally structured) organs and structures in many cases. I showed elsewhere how this principle of 'alternative matching' could account for a great variety of experimental data[21]. (It might be argued that in the case of Lorber's subjects most or all cells had been preserved *per se*, but were only squashed together, and that only post-mortem detailed examination could unambiguously settle this issue. Yet this seems highly improbable in view of the normal fragility of brain neurons under pressure. At any rate, such explanations could hardly account for the findings reported by Lenneberg and Lashley, cited earlier).

I conclude that proper brain functioning with anatomically abnormal brains (whose structuring is nevertheless still genetically guided) is perfectly consistent with normal cognitive and behavioural performances by these abnormally structured brains, as long as *some* of the right kinds of cells are still available. Because of redundancies of cell types in the normal, probably strongly genetically controlled, developmental programme, some nervous systems can dispense with a considerable number of cell types whose function can be taken over by the remaining ones, in a way explained by earlier publications[21a]. Thus, while evolution seems to have provided for high cell redundancy in the case of nervous systems of many species, thereby safeguarding these systems against accidental cell damage, and against the known continually occurring death of neurons (some 5×10^4 human adults neurons have been estimated to die per day), much of this redundancy is apparently as species-specific, and, hence, as genetically encoded, as the building plan of the organism.

If we concede that the developmental blue prints of the structure of human and animal nervous systems are strongly genetically determined, then this does not imply, at least for higher organisms (which are less stimulus-bound than lower ones[22]), that the performances of mature nervous systems depend exclusively on developmental (genetically programmed, though environmentally influenced) factors. Nervous systems are subjected to modifications caused by self-generated creative activity that involves learning, quite apart from normal learning induced by the environment. (Admittedly, nobody knows for sure what engrams consist of at the molecular biological level, although there is no shortage of conjectures[23]). Apart from this, ageing occurs. Thus, the innate-versus-environment controversy about behaviour is an artefact of unjustifiably opposed extremes. Neither the building plans of nervous systems (and the genetic control of these plans) nor modifications of these systems due to learning and ageing (etc.) are independent determinants of behaviour. The latter is a result of the developmentally generated structures *and* their life-long modifications due to the nervous system's activity and ageing and its encounter with the environment. Thus, the first question posed at the beginning of this paper turns out to be unanswerable, and unfit to be the subject of further serious debate. There are such a large number of genetic variables and environmental and developmental contingencies, and contingencies due to ageing of the brain's (etc.) machinery, at work, that a 'how much' genetic and 'how much' environmental control question becomes plainly absurd.

Those who seriously believe that they can demonstrate within the species *Homo sapiens* evidence of racial superiority, as regards intelligence, should remember the vast number variables that participate in producing the kind of behaviours that we regard as intelligent. Only a scientifically untutored simpleton would argue that intelligence is mainly genetically controlled. On the other hand extreme environmentalists are as unjustified in their views as extreme geneticists, since there can be no doubt that the machinery of behaviour is developed in each human individual under considerable genetic control. Yet, this machinery, even if optimally developed, does not necessarily ensure that its user makes optimal use of it. People whom we would regard by common sense intuition as 'very intelligent' may use their intelligence very suboptimally and turn out to be failures in life (perhaps also because of contingent circumstances), whereas others, suboptimally endowed might greatly enrich our culture by making maximal use of their 'inferior' brains. Who knows or can know?

II

I have already partially answered the second question posed at the start of the paper. Animal ethology, of course, provides ample examples of completely or partly stereotyped behaviour which is sometimes referred to as being 'innate'. A classical example which I am fond of quoting comes from Lorenz[24]. He writes:

> A young male salticid spider which, after moulting for the last time, approaches a female, must neither mistake another species for his own, nor

> must he perform the signalling of his courtship dance in any other way than the one to which his female responds; otherwise he would be eaten by her immediately. He has no opportunity in his short life to gain any information about what a female of his species looks like, nor what movements he must perform to inhibit her feeding reactions and to stimulate her specific mating responses.

Here, then, we have apparently pretty rigidly stimulus-bound behaviour, which appears to be strongly genetically programmed for the species. In such cases it looks suspiciously as if genetic control is very strong. It would seem that genes, by controlling the development of the nervous system, produce a nervous system in this species, which is extremely exact in its reaction to certain environmental stimuli, which have highly selective value.

Yet, from this example we can hardly conclude that, with the exception of some trivial examples mentioned by human ethologists, the human nervous system is as selective in its cues, and as specifically responsive to them, as that of the salticid spider. Nevertheless, we can be sure that people have strong food-preferences, sex-preferences and so forth. But a human does not have to perform coitus in a stereotyped fashion. Much existing sex literature suggests that this is not the case. Here man and the salticid spider differ drastically. Let us look at another striking example cited by Lorenz in his book (which was recently and may still be out of print).

> A young swift reared in a narrow cave in which it cannot extend its wings (far less beat them up and down), in which it cannot attain a sharp retinal image as the farthest point of the cave is much nearer than its shortest focusing distance, and in which it cannot gain experience on parallactic shifting of retinal images, nevertheless proves to be perfectly able on the very moment it leaves the nest cavity to assess distances by the parallactic shift of the object's images. It can also cope, in its rapid flight, with all the intricacies of air resistance, upcurrents, turbulence, and air pockets and can 'recognize' and catch prey, and finally effect a precise landing in a suitable place.

The last passage quoted suggests that the wired-in nervous system in this case must have arisen under genetic control so accurately as to endow it with the precise performance competences, and to adapt these competences, without training, to existing environmental circumstances. Here we have apparently a much greater degree of flexibility of the behaviour repertoire than in the salticid spider, although the behaviour competences as distinct from their detailed enactments seem to be genetically strongly determined. Just the same, because of the extent of variation between species in the degree of stimulus-binding we cannot generalize from one species to the next, let alone from the swift to man. Hence, the answer to the second question, raised at the beginning of the paper is as follows. In some cases complete behaviour modes seem to be largely or nearly completely determined genetically in most of their minutiae, and leave little room for behavioural adaptation. In other cases, the degree of genetic determination of behaviour modes may be very weak, and environmental givens may predominate in determining the behaviour. Many other cases may lie between these extremes. Hence, once more the request to decide between two extremes, which is often voiced, seems to be gratuitous.

III

I have already answered part of the third question, namely the extent to which man's intelligence depends on either the environment or on genetics. I must now try to deal with the issue of the genetic-versus-environmental dependence of morality. I shall first digress somewhat. Psychologists of long past generations have sometimes contrasted 'the intellect' and 'the emotions' a division which nowadays may seem very artificial. I have already stressed that intricate human cognitive activity is not stereotyped, as witnessed by the works of men and women of genius. We may, as noted, suppose that genes provided programmes for laying down the machinery for enacting the magnificent creations of these people of genius. But the intellectual products themselves depended on past learning and, probably, on brain-represented trial-and-error accompanied by psychological 'sets' that provided goals, and they also depend on value judgements and decisions[25], and so forth, thereby invoking the gamut of cognitive creative machinery.

When it comes to issues of morality some moral philosophers have partly misunderstood people like Dawkins who talk about 'selfish genes' in a purely technical sense of population genetics. Selfishness in the behavioural sense could be partly genetically and partly environmentally produced. Yet, when behaviour-orientated population geneticists and various animal sociobiologists talk about the 'selfishness' of genes, they use this term in the sense of technical jargon[26]. Dawkins' 'fundamental law of gene selfishness' involves enhancement within the population of those genetic systems that fix certain relevant allegedly 'innate' behaviour strategies, i.e. the genes concerned spread through the gene pool of the population and supposedly become fixed by favourable selection (at the level of individuals, and not by 'group selection' as Dawkins has stressed[27]). What becomes selected in this case is an innate behaviour programme, e.g. something as rigidly fixed as the behaviour of the salticid spider cited above. Hence, what is 'selfish' about the genes involved is simply that they may favour their owners when seeking out ecological niches that tend to select particular behaviours of these owners. Owners of selfish genes, by being behaviourally well adapted via these genes, tend to maintain these genes in the gene pool of the population. In this sense the 'selfishness' of these genes is no different from that of any other genes that can provide selective advantages for their owners in a suitable ecological niche.

Trouble starts when people try to extrapolate from 'selfish genes', that preserve, say, 'innate' animal behaviour capabilities of the simple (yet, nevertheless intricate) kind mentioned, to wholesale interpretations of control by 'selfish genes' in human populations, and when such genetic control cannot be analyzed independently. Superiority of 'Aryan genes' as claimed by many Nazis, suggested that these alleles lead to behaviour traits which made the behaviour of their carriers superior to that of the genetically less well-endowed 'Non-Aryans' (Jews, Negroes, etc.). The superior genes were then supposed, because of their stronger selfishness, to assert themselves in the gene pool of the population, thereby making the master race, whose members carried these more selfish genes, 'superior' and dominant over the genetically less well endowed people. Such extrapolations, jointly with the belief that only the genetically 'superior' have a right to exist, have led to the Nazi-instigated extermination of Jews and other supposedly 'inferior' people, and have inspired feelings of many racists, often in high positions in many countries.

Lewontin[28] has termed the tendency to extrapolate Darwinism to the extent of asserting that every discoverable human trait (notably every behavioural trait), is predominantly genetically determined, and has positive or negative selective value as far as its carriers are concerned, a vulgarization or a caricature of Darwinism. In fact, if a Jew built up a successful business which was strongly competitive with the enterprises of several non-Jews, then greedy and stupid Nazis and neo-Nazis would argue that the gene pool of the Jewish population is saturated with genetic alleles coding for 'money grabbing' traits and that the Aryan population must exert 'natural selection' against the Jewish population by exterminating its members and robbing their property. There is plenty of evidence of this kind of 'philosophy' if you wish to call it such.

Lewontin, in the review mentioned, writes that

> For more than forty years evolutionary theory has remained free of a naive selectionism, but in recent times there has been a return to the extreme form of the adaptationist program, as evolutionists have rediscovered behaviour. Beginning with the undoubted truth that behaviour must, like morphology and physiology, be subjected to the force of natural selection, the new

Panglossians end with the old error that *all* describable behaviour must be the direct product of natural selection.

Unwarranted extrapolations from the findings of ethologically orientated population geneticists are based on two assumptions, which figure implicitly in the philosophy of Dawkins. In the preface to his book[29] he wrote (a passage pounced upon by Lewontin)

> We are survival machines robot vehicles blindly programmed to preserve the selfish molecules known as genes. This is a truth which still fills me with astonishment.

The trouble with this kind of extremism is manifold. First, although people, as well as animals, may well be robots a view to which a materialist like myself basically subscribes[30] we are *not*, as I have stressed previously, demonstrably gene-dominated robots. (Whenever a criminal monopolist Nazi industrialist employed Jewish or Russian (etc.) slave labour, and worked these people to death in order to enrich himself, then this could hardly be attributed to the natural selection of his 'Jewish money-grabbing genes' since a Nazi was supposed to be a perfect Aryan free from such genetics. Possibly the money-making trend of certain people is acquired as a result of learning that money-making is a helpful trait. As Brecht put it in 'Die Dreigroschenoper', 'only those who live in a state of wealth live pleasantly' - undoubtedly not universally true either.)

Human beings are not simple survival machines. The old idea that a chicken is there to produce an egg for making another chicken is a somewhat ridiculous idea when applied to the chicken, and even more absurd when extrapolated to Man. Beethoven and Newton left no issue, but their creations have produced immense values. What counts in man is the creation of novel ideas and *values*, and not just fertility. Some artists and scientists, and other notable contributors to human culture and society, have undoubtedly produced sizeable families (e.g. J.S. Bach), but they are not primarily known for their fertility, but for their artistic produce, much of which depends on learning as well as on genetics. To make progress, nowadays, in say, a scientific theory such as quantum chromodynamics, one must first *learn* the existing theory, and any genetically programmed human brain is not equipped at birth with the details of quantum chromodynamics. At any rate, as noted earlier, what is generally wired-in, as regards human nervous systems, is, in the case of flexible cognitive behaviour, a set of propensities rather than detailed behaviour items. Elsewhere, in a speculative frame of mind, I suggested in detail how genes could control the wiring-in of brains[31] and, thereby provide the machinery for, say, cognitive functions, without implicating any genetic determination of the use to which that machinery is to be put in contingent circumstances. I think that one can be both a materialist and a reductionist[32] (but see Lewontin[33]) without embracing the view that reductionism requires that all uses that a machine is or can be put to are to be explained in terms of the machine itself.

Organisms could be machines, but they are not closed and isolated systems, and live in environments that contain myriads of contingencies, and reductionists only wish to explain the *principles* of structuring and functioning of organisms in terms of the properties of their material constituents at appropriate levels of reductive discourse, and *some* of these constituents happen to be genes. It is perfectly sound for a combined materialistic and reductionist enterprise to try to explain how organismic development, including the development of nervous systems, could come about as a result of genetic control *combined* with a good many other mechanistic building principles. The mechanistic and reductionist points of view seem to me mutually perfectly compatible. Materialists, like myself, claim that all events in the universe are consequences of properties of matter undoubtedly a metaphysical creed where conscious events are also to be conceived of as 'properties of matter[34]'. This very general metaphysical belief, however, does not automatically explain the precise ways in which classifiable phenomena, studied by scientists, are properties of matter. To explain these phenomena one requires hypothetico-deductive (and reductive) theories that are applicable to a host of mechanisms[35].

I conclude that although organisms could be machines in a sense that can be specified *via* hypothetico-deductive theories (combined with metaphysical assertations, such as I propounded elsewhere[36]), organisms are not simply survival machines. Even if genes are selfish in Dawkins' technical sense, then this does not make us mere survival machines, for reasons given. I must now turn to the relation between morality and genetics in the less technical sense.

As I wrote elsewhere[37], with luck science might discover *how* human values are represented by brains, but science cannot tell us which values we ought to adopt (e.g. which moral values). Science is not ethically normative, and with this position Dawkins seems to be in agreement. It seems to me quite feasible that some human values (such as 'goodness', 'wickedness', and so forth) might well be innate, and that the brain might represent such values by means of appropriate constructs (conceivably certain proteins[38] although this is pure speculation). Yet, even if some values are innate then this does not tell us which of these innate values becomes associated with brain representations of particular things or situations or actions etc. Such associations may, possibly in most or all cases, be based on learning. Some societies learn that cannibalism is a good thing. Other societies teach people that to kill others in war is, in specific circumstances, a good thing, and reward this by medals. I conclude that although values may be native the application of these values to life situations might, for all we know, be predominantly learning-dominated.

Moral philosophers are not just concerned with human values, but also with the emotions that are sometimes, possibly often, involved in value-saturated situations. Thus Mrs. Midgley[39] harps on a supposed 'disregard of emotional context' in Dawkins' discussions of animal behaviour. She takes it for granted that animals have emotions in the sense in which humans have them. A dog is supposed to have such emotions, according to some dog lovers, but does a sea urchin? Finding brain regions in a chimp that resemble those of humans, and showing that these human brain regions are associated with emotional behaviour (e.g. some hypothalamic regions) does not permit us to conclude that a chimp has emotions in the human sense. For all we know particular emotions could be related to specific states of certain types of proteins in particular neurons in the brain of man and chimps. Yet, the emotion-mediating protein types of the chimp might differ very substantially from those of man. Granted that the emotion-mediating proteins (if such proteins exist) would be coded for by genes, this does not permit us to infer that the types of emotions a chimp can form are type-by-type, or at least for some types, similar to certain human emotions. I conclude that extrapolations from human kinds of emotions or human kinds of values to any allegedly corresponding animal emotions or values is unjustifiable.

The advantage of 'being social', according to those moral philosophers who share Mrs. Midgley's point of view, is 'only possible as part of a whole complex way of life in which outgoing emotions which [philosophical] egoism denies constantly work for harmony.' Yet, it is hard to see how the notion of 'emotions constantly working for harmony' differs in its degree of drastic *ad hocness* from the optimization criteria of sociobiological minitheories of the kind that I discussed elsewhere[40], and which Mrs. Midgley regards with contempt (e.g. the game-theoretical approaches to 'evolutionary stable strategies' (ESSs) as applied to animals). It seems hard enough to guess human motives, and human emotions, on the basis of wild *ad hoc* assumptions, let alone compare human motivations and alleged animal motivations. Indeed, those who are highly sceptical of Freudian psychodynamic theory have often noted that with the help of sufficient *ad hoc* assumptions almost any human behaviour can be interpreted in terms of Freudian motivational theory. Those who are analogizing from man to beast or *vice versa*, particularly where motivations or emotions or moral values are concerned should bear in mind the complete *ad hocness* of this enterprise.

Comparing this *ad hocness* with Freudian theorizing seems to me very appropriate. In this context an example given by Frankl[41] may be helpful. He referred to

> ...a famous Freudian analyst [who] has recently published a book on Goethe. Let me quote from the review of the book by Heuscher. 'In the 1538 pages the author portrays to us a genius with the earmarks of a manic-depressive, paranoid and epileptoid disorder, of homosexuality, incest, voyeurism, exhibitionism, fetishism, impotence, narcissism, obsessive compulsive neurosis, hysteria, megalomania, and so forth. The author seems to focus almost exclusively upon the instinctual dynamic forces that underlie the artistic products. We are led to believe that Goethe's work is but the result of pre-genital fixations. Goethe's struggle does not really aim for an ideal, for beauty, for values, but for overcoming an embarrassing problem of premature ejaculation. These volumes show us again' the author of the review concludes 'that the basic position of psychoanalysis has not really changed.'

To sum up, then, Mrs Midgley's analogizing about animal behaviour and human behaviour strikes me as being no more justified than the Freudian *ad hocery* just quoted. Also Dawkin's view of man as a gene-dominated robot lacks justification. Biology might, in due course, lead us to detect many genes that control behaviour. Yet this, *per se,* does not permit us to infer which particular genes are involved in any idiosyncratic mode of human behaviour. At best we might be able to detect genes for those types of strongly stimulus-bound behaviour exhibited by, say, the mating behaviour of the salticid spider discussed above. But none of this permits us at present to make sweeping generalizations about the sociobiology of man.

University of Newcastle upon Tyne NE1 7RU, England

Notes

1 Dawkins, R. *The Selfish Gene*, London, Oxford University Press (1976).

2 Mackie, J.L. 'The Law of the Jungle: Moral Alternatives and Principle of Evolution', *Philosophy* (1978), **53**, 455-464.

3 Wassermann, G.D.'A Critique of some Aspects of Human Ethology', *Behavioural and Brain Sciences* (1980), **3**, 630-631. (See Chapter 8 above)

4 Midgley, M. 'Gene-juggling', *Philosophy* (1979), **54**, 439-458.

5 Op. cited note 1.

6 Parker, G.A. 'Selfish Genes, Evolutionary Games, and the Adaptiveness of Behaviour', *Nature* (1978), **274**, 849-855.

7 Skinner, B.F. *Science and Human Behaviour*, New York, Macmillan, 1953.

8 Flynn, J.R. *Race, IQ and Jensen*, London, Routledge and Kegen Paul, 1978.
9 Lynn, R. 'IQ the Environmentalist Position', *Nature* (1981), **291**, 95; J. Beloff (ed) *A Balance Sheet on Burt*. Supplement to the Bulletin of the British Psychological Society, vol. 33, London, British Psychological Society, 1980; Thoday, J.M. 'Probity in Science: the Case of Cyril Burt', *Nature* (1981), **291**, 517-518.
10 Feldman, M.W. and Lewontin, R.C. 'The Heritability Hang-up', *Science* (1975), **190**, 1163-1168.
11 Op. cit. note 3.
12 Cf. W. Penfield and T. Rasmussen, *The Cerebral Cortex of Man: A Clinical Study of Localization of Function*, New York, Macmillan, 1950; C. Blakemore, *Mechanics of the Mind*, Cambridge, Cambridge University Press, 1977, pp. 99-101.
13 Galau, G.A., Klein, W.H., Davis, M.M., Wold, B.J., Britten, R.J. and Davidson, E.H. 'Structural Gene Sets Active in Embryos and Adult Tissues of the Sea Urchin' *Cell* (1976), **7**, 487. (Structural genes code via messenger RNA for polypeptide chains of proteins).
14 Wassermann, G.D. *Molecular Control of Cell Differentiation and Morphogenesis*, New York, Marcel Dekker, 1972; Wassermann, G.D. 'Molecular Genetics and Developmental Biology', *Nature New Biology* (1973), **245**, 163-165; Wassermann, G.D. *Neurobiological Theory of Psychological Phenomena*, London, Macmillan, 1978.
15 Jones, J.S. 'How much Genetic Variation?' *Nature* (1980), **288**, 10-11.
16 Lenneberg, E.H. in *The Neurosciences, Second Study Program*, F.O. Schmitt, ed. New York, Rockefeller University Press, 1970, p. 361.
17 Lorber, J. 'The Disposable Cortex', *Psychology Today* (1981), **15**, 126.
18 Lashley, K.S. 'In Search of the Engram' *Symposia of the Society for Experimental Biology* (1950), **4**, 454-482.
19 Op. cited note 14 (1978).
20 Zipser, B. and Mackay, R. 'Monoclonal antibodies distinguish identifiable neurones in the leech' *Nature* (1981), **289**, 549-555 and Op. cited note 14 (1978).
21 Op. cited note 14 (1972) and (1978).
21a Op. cited note 21 (1978).
22 Hebb, D.O. 'The Problem of Consciousness and Introspection' in *Brain Mechanisms and Consciousness* (J.F. Delafresnaye ed.)(Oxford: Blackwell, 1954), pp. 402-441, see his p.406.
23 Op. cited note 14 (1978).
24 Lorenz, K.Z. *Evolution and Modification of Behaviour*, London, Methuen, 1965.

25 Cf. Wassermann, G.D., Review article on *The Biological Origins of Human Values* by G.E. Pugh (Routledge and Kegan Paul, 1978), *Philosophia* (1982), **11**, 407-414.

26 Cf. op. cited notes 1 and 6.

27 Op. cited notes 1 and 2.

28 Lewontin, R.C. 'Caricature of Darwinism', *Nature* (1977), **266**, 283-284.

29 Op. cited note 1.

30 Wassermann, G.D. *Brains and Reasoning*, London, Macmillan, 1974; Wassermann, G.D. 'Reply to Popper's Attack on epiphenomenalism' *Mind* (1979), **88**, 572-575; Wassermann, G.D. 'Materialism and Mentality' *Review of Metaphysics* (1982), **35**, 715-729.

31 Op. cited note 14 (1978) and Wassermann, G.D. 'A New Type of Neuropsychological Theory' in *Cognition and Brain Theory* (1980), **3**, 33-36.

32 Cf. Op. cited note 30 (1974).

33 Op. cited note 28.

34 Op. cited note 30 (1979) and (1982).

35 Op. cited note 30 (1974); Wassermann, G.D. 'On the Nature of the Theory of Evolution', *Philosophy of Science* (1981), **48**, 416-437; Wassermann, G.D. 'On the Laws of Nature', *Synthese* (1982), **51**, 381-396; Wassermann, G.D. 'Biological Aspects of the Philosophy of Mind', *Metaphilosophy* (1980), **11**, 199-209.

36 Op. cited note 30 (1982).

37 Op. cited note 25

38 Op. cited note 14 (1978).

39 Op. cited note 4.

40 Op. cited note 35 (1981).

41 Frankl, V.E. 'Reductionism and Nihilism' in *Beyond Reductionism* (A. Koestler and J.R. Smythies, eds.), London, Hutchinson (1972), pp. 396-408.

13 Neuropsychology vis-à-vis Skinner's behaviouristic psychology

Skinner's approach to the differences and similarities of (learning-acquired) ontogeny and (predominantly genetically controlled) phylogeny of behaviour follows the holistic tradition of stimulus-response (SR) behaviourist theorizing, which is also adopted by ethologists. This theorizing assumes that the organism can be treated like a black box, whose stimulus inputs can be related to its response outputs. Inside the black box reinforcement and partly genetically controlled processes occur, as appropriate. Yet SR theory makes no attempt to explain either ontogenic or phylogenic aspects of behaviour in terms of the machinery of the black box. At most there are oblique hints, as when Skinner states 'But we must look to the construction of the machine, as we look to the phylogeny and ontogeny of behaviour, to account for the fact that an ongoing system acts as if it has a purpose.'

The philosophy of SR theory, whether dealing with the ontogeny or the phylogeny of behaviour, is based on the belief that a few 'molar variables' can suffice to formulate complex behaviour. This outlook is closely analogous to the belief of some bygone generations of physicists who thought that simple physical relationships such as Ohm's law, or van der Waal's equation of state for gases, were paradigmatic for all physics. Indeed, statistical mechanics, by averaging over the energy states of an ensemble of gas molecules, was able to derive equations of state of reasonably simple kinds. Yet it must be stressed that a gas, or for that matter simple solids (such as a piece of metal), are compositionally relatively uniform compared to a central nervous system (say of a mammal) which, at the molecular level, is probably intercellularly extremely heterogeneous (Wassermann 1978). It is for this reason that one cannot apply statistical mechanics (as distinct from thermodynamics) to a complex multineuronal system. Instead, many people have come to recognize that cognitive processes of man depend on a host of parameters, whose effects cannot be expressed in terms of simple laws with the help of statistical mechanics (and thermodynamics) or otherwise.

To invoke 'reinforcement' in the ontogeny of behaviour, as frequently as Skinner prefers, seems to me not unlike the wholesale invocation of 'natural selection' to explain almost any facet of evolutionary adaptation, as practised by panselectionists (cf. Dennett: 'Intentional Systems in Cognitive Ethology' *BBS* 6(3), 1983), or like attributing all this to divine creation as is done by creationists. Realizing this, one group of scientists has abandoned realism, for the time being, and has resorted to (usually) computer-simulated formulations of cognitive processes (the simulating system being the computer program, i.e. its software, and not the computer hardware) in terms of artificial intelligence (AI) systems. Other theorists (including myself) have tried to relate the functions of nervous systems and behaviour more directly in terms of neuropsychological theories (cf. Hebb 1949; Sommerhoff 1974; Wassermann 1978, particularly chap. 6, for discussions of various types of neuropsychological theories, including my own). In contrast to various theoretical neuropsychologists and AI theorists, Skinner's SR theorizing, as he admits, 'leaves out of account habits, ideas, cognitive processes, needs, drives, traits, and so on. But it does not neglect the facts upon which these concepts are based'. To some extent this applies also to those neuropsychologists who, like Skinner, try to rid themselves of 'mentalistic ghosts' without ignoring the facts related to these ghosts, and, who, like myself, are hard-faced materialists. It is therefore worthwhile to compare Skinner's approach with that of the latter group of neuropsychological theorists.

Although Skinner does not assume that the nervous system (NS) is a tabula rasa at birth, he leaves aside the high degree of developmental organization of the NS, which seems to be largely gene dominated (even if environmental factors determine ultimate details). Jacobson (1969, p. 543) remarked pertinently that in *normal* embryonic development

> developing neurons sprout slender processes, their axons and dendrites, which in some cases grow to relatively great lengths to form connections with other neurons. The direction of growth of these processes and the targets on which they terminate appear to be constant in all individuals of the same species. Anatomical and physiological methods have shown the remarkable invariance of neuronal circuits and have given no evidence of random connectivity. A distinguished neuroanatomist (Palay, 1967) has recently written, 'The nervous is not a random net. Its units are not redundant. Its organization is highly specific, not merely in terms of the connections between particular neurons, but also in terms of number, style, and location of terminals upon different parts of the same cell and the precise distribution of terminals arising from each cell.'

Elsewhere I have written (Wassermann 1978, p. 33) 'Perkel (1970) also noted that "increasing evidence for the homology of neurons identifiable from individual to individual (Bullock 1970) bespeaks a more basic role for 'wired-in' neural connections than has been attributed to them." ' Such observations led me to conclude that neural nets are highly species-specifically wired-in under predominant genetic guidance (at least in *normal*, i.e. nonexperimental, development). This suggests that learned behaviour (LB) as well as phylogenetically derived ('innate') behaviour (PB) can be interpreted as due, in part, to modifications of essentially genetically wired-in nervous systems. Modifications of NSs related to either LB or PB could be similar in kind, except that the modifications related to PB could become developmentally established under genetic control. If one deals with similar *types* of modifications of NSs in PBs and LBs, then it is not surprising to discover the numerous parallels between LBs and PBs which Skinner has elegantly presented (explicitly or implicitly).

Although a holistic analysis of Skinner's type is valuable in its own right, it has its limits in that it only elucidates what (quasi black box) organisms do in SR situations. It does not tell us anything, however, about the mechanisms that are involved in PBs and LBs. Behavioural science, if it is to be ranked as a science, must also be committed to the discovery of mechanisms, notably nervous system mechanisms at the physiological and molecular biological levels. The latter levels ultimately link with genes (Wassermann 1978) and, hence, with evolution. Granted that 'reinforcers' play a role in LB, we must understand their neural equivalents. Moreover, we can be sure that LBs and PBs of all kinds, whether occurring in simple SR chains or in complex cognitive behaviour, are hierarchically organized into units. I suggested elsewhere that cognitive hierarchies could *inter alia* comprise hierarchies of concept-representing units (Wassermann 1978). If so, what is wired into NSs are possibly genetically provided macromolecular concept-representing units (CRUs) rather than specific concepts (with a few possible exceptions). Learning processes could then determine, on the basis of contingencies, to which genetically established CRU a particular (learned) concept becomes allocated. This is where Skinner's environmental contingencies become relevant. CRUs could form a hierarchy of associated systems (Wassermann 1978). It seems to me that although reinforcement processes are likely to play an important role in strengthening genetically established connections between CRUs, ultimately extensive neuropsychological research will be required in this area. (It is not good enough to state like Skinner, at the end of his section 'Interrelations among Phylogenic and Ontogenic Variables' that 'the behavior of civilized people shows the extent to which environmental variables may mask an inherited endowment.' *How much* behaviour is inherited and how much learned seems to me to be a practically insoluble problem; cf. chapter 12).

Long ago Hebb (1949) clearly recognized, like others before him, that cognitive structures such as 'sets', although learnable (Allport 1955), are represented within NSs, and SR talk cannot explain this away. Moreover, many 'sets' are goal directed, and how goal direction is neurally (or intraneurally?) represented remains an unresolved problem (*pace* Sommerhoff 1974). Indeed, notwithstanding detailed analyses of teleology (MacKay 1951; 1952; Sommerhoff 1950; Wassermann 1981; Woodfield 1976) there exists no agreement on how to define 'goals'. It seems to me that Skinner's discussion of 'sense of purpose' (in his final section) or 'intention' disregards the fact that goal-directed behaviour of people or animals is probably more than an expression of 'operant reinforcements'. At least this assertion remains tenable until the nature of goal directedness and of 'operant reinforcement' has been resolved in terms of experimental and theoretical neuropsychology. I conclude that Skinner's type of comparative analysis, although molar, is an important step in the right direction. His type of analysis must ultimately be extended to the level of neural mechanisms, where the relation between the ontogenesis and the phylogeny of behaviour can become more fully elucidated. I am not advocating reductionism for its own sake, but recommend it in the belief that it may afford a deeper level of understanding of various aspects of behaviour than unsupplemented holistic analyses.

Bibliography

Allport, F.H. (1955) *Theories of perception and the concept of structure*, New York, John Wiley.

Bullock, T.H. (1970) *The reliability of Neurones* Jacques Loeb memorial lectures, Woods Hole, Woods Hole Oceanographic Institution, Massachusetts.

Hebb, D.O. (1949) *The organisation of behaviour*, Wiley.

Jacobson, M. (1969) 'Development of specific neuronal connections', *Science*, **163**, 543-47.

MacKay, D.M. (1951) 'Mindlike behaviour in artifacts', *British Journal for the Philosophy of Science*, **2**, 105-121.

MacKay, D.M. (1952) 'Mentality in machines' *Proceedings of the Aristotelian Society* Supplem., **26**, 61-86.

Maynard Smith, J. (1974) 'The theory of genes and the evolution of animal conflict', *Journal of Theoretical Biology*, **47**, 209-221.

Maynard Smith, J. (1976) 'Group Selection', *Quarterly Review of Biology*, **51**, 20-30.

Maynard Smith, J. (1978) 'Constraints on human behaviour', *Nature*, **276**, 120-121.

Palay, S.L. (1967) 'Principles of cellular organisation'. In *The Neurosciences* ed. G. C. Quarton, T. Melnechuk and F.O. Schmitt. New York, Rockefeller University Press.

Parker, G.A. (1978) 'Selfish genes, evolutionary games and the adaptiveness of behaviour', *Nature*, **274**, 849-855.

Perkel. D.H. (1970) 'Spike trains as carriers of information,' in *The neurosciences second study program*, ed. F. O. Schmitt, Rockefeller University Press.

Rose, S. (1978a) Essay review of E.O. Wilson's *On human Nature, Race and Class*, **20**, 277-287.

Rose, S. (1978b) Review of E.O. Wilson's *On human Nature, New Scientist*, 19th October.

Skinner, B.F. (1984) 'The phylogeny and ontogeny of behavior', *The Behavioral and Brain Sciences*, **7**, 669-711.

Sommerhoff, G. (1950) *Analytic biology*, Oxford, Oxford University Press.

Sommerhoff, G. (1974) *Logic of the living brain*, Wiley.

Wassermann, G.D. (1972) *Molecular control of cell differentiation and morphogenesis*, New York, Marcel Dekker.

Wassermann, G.D. (1973) 'Molecular genetics and developmental biology', *Nature New Biology*, **245**, 163-165.

Wassermann, G.D. (1978) *Neurobiological theory of psychological phenomena*, London, Macmillan and Baltimore, University Park Press.

Wassermann, G.D. (1981) Book review of *Teleology* by Andrew Woodfield, *Philosophia*, **40**, 125-32.

Wilson, E.O. (1975) *Sociobiology: the new synthesis*, Cambridge Mass., Belknap Press of Harvard University Press.

Wilson, E.O. (1978) *On human nature*, Cambridge Mass., Harvard University Press.

Woodfield, A. (1976) *Teleology*, Cambridge, Cambridge University Press.

14 Social aspects of rationality

14.1 Introduction

This paper has three aims. First it tries to re-examine the notion of rationality. Second, it will examine the criteria by which the supposed presence or absence of rationality in individuals or societies are adjudicated. Third, it attempts to debate the means whereby these criteria are defended, rejected and/or replaced. Finally it will be suggested that a suitable neuropsychological theory could provide a *metatheory* for understanding how biological machinery could provide for the means of attaining rationality, thereby suggesting that materialism could be consistent with rationality. There exists a voluminous literature on each of the first three topics, and I shall, as I go along, only refer to a few sources which, however, give, when consulted, access to a rich store of references on various issues linked with rationality[1]. *I shall primarily defend a relativistic view of rationality.*

14.2 Status and claimed relativity of rationality

If we look at any contemporary, reasonably comprehensive, work on theoretical cognitive psychology[2] we find that problem solving, decision making, value formation (and related aesthetic activities), creativity of the most varied kinds (such as writing novels, composing music, inventing new scientific theories, or new technological processes etc.) form prominent aspects of the typical complex cognitive 'repertoire'. Many of the varied cognitive activities, particularly the more intricate ones, can be performed intelligently or unintelligently, although perhaps not too many people are agreed on what we mean by 'intelligence'. Some will say that intelligence is what an IQ test measures, others will strongly dispute this, and some may adopt some widely current definition like that of W. Stern, who according to Slater and Roth[3] defined intelligence as 'the general ability to adapt to new situations by means of purposeful thinking'. Yet, this definition of intelligence only shifts us to the request for a definition of 'purposeful thinking' thus leading to the central problems of teleology, some of which I discussed elsewhere[4]. At any rate, intelligence does not amount to rationality. Cognitive procedures can be carried out intelligently, yet they may be quite irrational. One only has to look at the case history of a serial killer, such as Peter Sutcliffe (the so-called 'Yorkshire Ripper') who is said to have suffered from paranoid schizophrenia and who murdered more than ten women quite intentionally. Or, on a vastly more frightening scale, there is the example of Adolf Eichmann, who deliberately instigated the organization of the murder of six million Jews during the second world war. Sutcliffe's and Eichmann's inhumanities were based on much 'purposeful thinking' and were 'well adapted to new situations', yet *most* of mankind would regard these perverted actions and the involved cognitive operations as being irrational. So, rationality is something distinct from intelligence, since intelligent people can be quite irrational at times (or in the case of Sutcliffe and Eichmann much of the time).

Some philosophers have attempted to formalize the notion of rationality, and they distinguish between 'theoretical rationality' and 'practical rationality'. According to Whiteley:[5]

> theoretical rationality involves four distinguishable operations: (1) apprising oneself of the data, (2) framing hypotheses to account for the data, (3) deducing the consequences of these hypotheses, and (4) estimating the relative probabilities of these hypotheses.

Whiteley's (1) to (3) are, most certainly, prominently involved in scientific theory construction[6], showing, not surprisingly, that well-formed scientific theories *are* partly rational in Whiteley's sense. At what level of probability is one to regard a hypothesis as being rational? Moreover, when assessing the acceptability of any hypothesis in everyday life (e.g. of the possibility that pigs can fly) people do not explicitly quantify the relative probability of the validity of that hypothesis. There exist, undoubtedly, many neuropsychologists and numerous artificial intelligence (AI) designers, who believe that neurons encode information on a probabilistic basis[7]. This, however, does not imply that all or any of the contents encoded by brains is *per se* of a probabilistic nature. There is no obvious necessity that the brain representation of, say, the smell of a rose or of a sensation of pleasure, has to be probabilistic. Even within the philosophy of science probabilistic approaches to the acceptability (or degree of confirmation) of scientific hypotheses have often been rejected. For instance, the suggested solution of the problem of scientific induction *via* Carnap's logical empiricism, which is based on the probabilistic evaluation of the validity of hypotheses, has been rejected repeatedly[8].

As regards theoretical rationality, as defined above, the proffered definition may be questioned, as being too restricted. For instance, decisions may form an important aspect of rationality, but they are not necessarily related to hypotheses. It could be argued that decisions form part of 'practical reasoning' (as defined below), but this is far from obvious. Likewise, there seems to be no obvious accommodation of value systems and aesthetics within the proposed framework of 'theoretical rationality'.

Let me now turn to the notion of 'practical rationality' as formulated by Whiteley[9]. He writes

> Practical reasoning may be analysed into stages or elements parallel to those discernible in theoretical reasoning. It involves (1) taking note of the facts of the situation, (2) framing a plan of action, (3) working out the (probable) consequences of adopting this plan, the means needed to carry it out and the further effects of fulfilling it, and (4) comparing for acceptability various alternative plans.

Whiteley asserts that '(1), (2) and (3) are analogous to their counterparts in theoretical reasoning' (see above) and he explains this at some length. He also argued that stage (4) of practical reasoning

> determines the choice between different plans, as stage (4) of theoretical reasoning determines the choice between different hypotheses. But here there is a great difference: for the choice is made not with reference to probability but with reference to satisfactoriness. What corresponds here to degrees of probability is the relative importance of our various wants.

Making a choice amounts, of course, to making a decision, but the converse is not true, not all decisions are based on choices. If I decide to have lunch at 1 pm, then I am not necessarily comparing this plan with various alternatives, such as having lunch at 2pm. Many of our decisions simply do not involve, in any form known to us, choices between alternatives. It could be argued, of course, by any psychoanalyst or kindred folk,[10] that any decision involves 'unconscious choices'. If one accepts arguments along that line one introduces, in my opinion redundant, metaphysics.

Whiteley also asserts (op. cited p. 82) that

> A man is practically rational in so far as he does what on the data available to him is likely to fulfil his settled and persistent intentions. He may be practically irrational either because of his judgement of the consequences

of his actions is faulty (a failure of intelligence) or he does what he realizes to be contrary to his settled and persistent intentions (a failure of will). A man whose beliefs and actions are determined by his faulty judgements is still in a sense rational.

Fulfilling settled and persistent intentions does not imply - in my opinion - that these intentions are rational, even if they are intelligently arrived at. This should be clear from the earlier cited examples of the mass murderers Sutcliffe and Eichmann, whose intentions were irrational, although settled and persistent and, alas, fulfilled. I conclude that Whiteley's definitions of theoretical and practical rationality are unsatisfactory. In fact, for reasons that should become increasingly clear as I proceed, I believe that any formal definition of rationality seems unobtainable. To see this let us look at psychiatry, where we see a supposedly rational psychiatrist often dealing with an apparently irrational patient. Let us ask what distinguishes the two, and whether we can get any clarification from the apparent, and often obvious, contrast. I shall mainly concentrate on schizophrenias.

Chronic schizophrenias of varied types are apparently linked to some brain abnormalities based partly on faulty genetics,[11] which suggests *inter alia* that (in some sense) 'normal' brains are essential for rationality. Yet, the normality of brains (even if it could be defined unambiguously, which seems highly unlikely) is not a *sufficient* condition for rationality. There are many examples of people who established and proved some of the most sophisticated mathematical theorems, but who, at times, behaved quite irrationally. (For instance it is told that one famous mathematician squashed the skull of another man with a tennis racket.) Even if we accept that brains develop largely subject to genetic control[12] then this genetic control may only ensure the correct wiring-in of brains, and may also be instrumental in the partial control of so-called 'innate behaviour' modes studied by numerous ethologists mainly in lower species (although there exist some prominent human ethologists). Yet, such genetic control of the development of the machinery does not guarantee that human brains are necessarily used rationally, since, as argued above, people capable of astonishing acts of rationality are equally capable of quite irrational thoughts and behaviour. This negative conclusion is not a specific consequence of studies of schizophrenia or of neuropsychology in general, but it seems to follow from everyday observations. Let us see whether the irrationality of schizophrenics leads to any special conclusions, which are relevant to the notion of rationality.

Some delusional thinking in schizophrenia may be directly or indirectly subsumed under von Domarus' principle[13] which in Arieti's formulation[14] asserts that

> whereas the normal person accepts identity only upon the basis of identical subject, the [schizophrenic] paleologician accepts identity based upon identical predicates.

As an example Arieti (op. cited p. 230) noted that

> A patient thought that she was the Virgin Mary. Her thought process was the following. 'The Virgin Mary was a virgin; I am a virgin, therefore I am the Virgin Mary'. The delusional conclusion was reached because the identity of the predicate of the two premises (the state of being a virgin) made the patient accept the identity of the two subjects (the Virgin Mary and the patient).

Other examples which show the operations of von Domarus' rule in schizophrenic thinking can be found in Arieti's monograph. It might be tempting to think that irrationality and rationality can be characterized by typically disordered logical thinking in the one case and 'correct' logical thinking of the 'rational person' in the other case. This, however, is not possible in any obvious manner. Significantly, Arieti[15] remarked that

> As a matter of fact anthropologic studies may disclose to the careful reader how often the principle of Von Domarus is applied in primitive thinking. Numerous studies, outstanding among which is the one by Storch,[16] have emphasized the similarities between primitive and schizophrenic thought, but the common underlying principles of logic that rule this thought have received no mention. Werner writes:
>
> It is one of the most important tasks of developmental psychology to show that the advanced form of thinking characteristic of Western civilization is only one form among many, and that more primitive forms are not so much lacking in logic as based on logic of a different kind. The premise of Aristotelian logic that, when a thing is *A* it cannot at the same time be *B*, will not hold true for the primitive...
>
> A Congo native says to a European: 'During the day you drank palm wine with a man, unaware that in him there was an evil spirit. In the evening you heard a crocodile devouring some poor fellow. A wildcat, during the night, ate up all your chickens. Now, the man with whom you drank, the crocodile who ate a man, and the wildcat are all one and the same person'. (Werner[17], Congo incident quoted from Lévy-Bruhl).

Here we see that a form of logic which may be subsumed under Von Domarus' principle[18] appears as irrational to members of Western societies, but apparently perfectly rational to Congo natives. This suggests strongly that one society's rationality may be another society's irrationality, so that the *form* of rationality seems to be society-bound. We have been led to this, far from obvious, conclusion by comparison of (to our society appearing as irrational) schizophrenics with supposedly 'normal' members of other societies.[18a]

This allegation of a strong relativism of rationality asserts that people can only *think* that they, or others, are rational, and that this thinking amounts to a belief or hypothesis, which does not imply that they or others are factually rational. Thus, from this point of view, rationality is not a fact but a hypothesis. Accordingly, since I claim that any assertion that something *is* rational is misguided, and that people *only can think* that something is rational, I am not forced to enter into a debate about the factual versus hypothetical basis of rationality, since I wish to repudiate the former. Because hypotheses cannot be shown to be true, but can only be falsified, the hypothetical rationality of certain people, whether claimed by them or others, cannot be shown to be absolutely true, and can only be a matter of consensus.

The similarity of schizophrenic and primitive thinking does not stop with the above comparison. For instance, hierarchically higher (i.e. more complex and abstract) levels of concepts must be *accessible* to 'normal' members of a sufficiently sophisticated society. Such hierarchically higher order concepts become apparently inaccessible to many types of schizophrenics, possibly owing to access blockage of certain parts of brains that represent these higher concepts in terms of biological machinery (e.g. if higher and lower order concepts are represented by different classes of brain components). This diminished accessibility of hierarchically higher order parts of abnormal brains could readily account for some well known clinical observations described in numerous texts on schizophrenia.[19] Typically it has been stated[20] that

> Schizophrenics are often very concrete and literal and lose the ability to think abstractly, to generalize or to 'get to the point' of figurative statements, such as proverbs, and have much difficulty in separating the relevant from the irrelevant... The patient seems to have his own elaborate and consistent but idiosyncratic rules of logic and language. Language is often stiff, stilted, inflexible, repetitious, and even stereotyped. Other features of schizophrenic language include neologism and incoherent 'word salad' in the most extreme form, though these can occur in manic toxic psychoses as well.

Notably the classic writings of Goldstein[21] illustrate that the thinking of certain classes of schizophrenics is more concrete than that of normal controls. Again, more primitive societies may show more concrete thinking than more advanced societies.

These, and some of the other facts mentioned suggest that the complexity of reasoning, as regards logic and conceptualization, is partly a matter of learning, and depends not only on the society to which people belong, but also on the educational contingencies to which any particular individual within a given society is exposed. Natural and artificial languages (such as mathematics and music scores) are among the most sophisticated means of expressing reasoning, and language is acquired, even if, as Chomsky and his followers believe, the deep structures of grammar may be innate.[22] For reasons similar to those given elsewhere in connection with the notion of 'intelligence' it seems to me unprofitable to debate how much of our capacity for reasoning is innate and how much is due to learning[23]. All that we can conclude at this stage is that 'rationality', if understood as the ability of individuals 'to reason in ways acceptably meaningful and plausible to others' depends on who the others are. A psychiatrist may not be able to make anything of the word salad of a schizophrenic, whereas an anthropologist educated in a sophisticated society may grasp the ways of reasoning of a member of a primitive society. Yet, this anthropologist cannot criticize the (to him alien) ways of reasoning of a more primitive thinker of a society other than his own as being 'irrational', whereas a psychiatrist can label schizophrenic thinking in members of his own society as 'irrational'. In other words there can be no fixed criteria of rationality which apply to all societies. Likewise, even within any given society there can be no norm. Suppose that most members of some primitive society reason by von Domarus' rule, but that there are a few exceptional members of that society who had received a sophisticated education in a more advanced society. Then these few exceptions would appear as 'irrational' to the majority of their own society, but might appear rational to members of more advanced societies.

If rationality amounts to 'acceptably meaningful plausible reasoning', then rationality depends, as is commonly observed, significantly on the individual standards of rationality adopted by various members of a given society, and these standards could vary considerably interindividually, so that, say, a politician who appears rational to his party may appear as 'irrational' to members of an opposition party. In the light of these conclusions let me cite some very different views of rationality, which are more in line with those of Whiteley's criteria, criticized earlier in this paper. Kekes[24] asserted that 'Potentially true or false beliefs are rational, while irrational beliefs lack this potential'. This assertion, however, disregards the fact that beliefs that appear as potentially true (or as actually true) and rational to a schizophrenic (e.g. that she is the Virgin Mary) or as actually true and rational to a member of the primitive society (e.g. the example cited by Werner above) may appear as being not only false, but irrational to other members of his own society (in the case of a schizophrenic) or respectively to members of other (e.g. non-primitive) societies. Thus the potential for truth or falsehood of propositions is, *per se*, insufficient for pronouncing ways of reasoning as rational. What may count is not just the potential truth or falsehood of beliefs, but the procedures of reasoning adopted in establishing some of these beliefs. For example, in establishing a hypothetico-deductive theory in some branch of science, the basic hypotheses cannot be shown to be true, but can (if the theory is good) be potentially falsified in Popper's sense. Yet, this alone does not remotely guarantee the 'rationality' of the basic hypotheses. Indeed, some apparently absurd hypotheses may be very suitable candidates for falsifications (e.g. the hypothesis that the earth is flat, or the hypotheses of astrology) and can also widely be accepted as irrational. Nevertheless, the fact that an individual of calibre, like C.G. Jung[25] tried to establish a case for astrology, provides further evidence that standards of 'rationality' vary intersubjectively. One man's rationality may be another's irrationality, notwithstanding frequent wide intrasocial agreement on what appears to be rational and irrational (see below). Moreover, the rationality of beliefs, within a society, may vary during the course of history. Anyone who nowadays believes in the validity of the 'atomic theory' advocated by Lucretius would be labelled an irrational crank.

The ultimate custodians of rationality are human brains (or conceivably other brain-associated matter systems), according to materialistic views (cf. op. cited, 1982 note 28 below). Since the output from any one particular human brain can sometimes be rational and at other times irrational it is *not* possible to suggest *sufficient* neuropsychological mechanisms for attaining rationality relative to any given specified standards. Likewise, it is not possible to suggest necessary neuropsychological mechanisms for attaining rationality, since theories of mechanisms depend on hypotheses, and hypotheses can never be shown to be necessary, but at best to be sufficient. Nevertheless, some people have already begun to suggest neuropsychological theories for rational processes in general. (Cf. op. cited note 7 for an explicit attempt in this direction). Such theories do not pretend to explain how particular individuals could be rational, but they try to explain how particular procedures of 'acceptably meaningful plausible reasoning' could in principle be enacted by postulated classes of brain mechanisms. Nobody who establishes such theories of brain mechanisms wishes to suggest that such general mechanisms are sufficient for explaining the rationality of particular individuals, since *idiosyncratic* reasoning or actions have to be explained, if at all possible, by further resort to *ad hoc* hypotheses. Similar remarks apply *mutatis mutandis* to the theories of all scientific mechanisms.[26] Hence, neuropsychology could certainly cast much light on theories of brain mechanisms capable of producing the kinds of behaviour that are widely accepted as rational by sufficiently advanced human societies. For instance, proving mathematical theorems represents a central rational activity.

It might be objected[27] that

> ...discovering how the working of the brain enables us to do mathematics will not illuminate the nature of mathematics: to understand that we must understand the relationship between numbers, how mathematical proofs work, and the standards by which we judge the correctness of a mathematical calculation or proof.

This argument, however, only raises a question, which, if answered, soon leads back to conclusions already reached about rationality. It may be asked 'In what sense do we 'understand' a mathematical proof, or any other aspects of mathematics?' If I question somebody about various aspects of a mathematical proof in order to judge his degree of understanding of that proof, then I must already understand the nature of that proof in order to assess the validity of his answers. So what does my own 'understanding' consist of? One familiar way of illuminating one's 'understanding' of, say, mathematical axioms depends on adopting the ways of linguistic philosophers. We could say that 'understanding' of axioms of mathematics is displayed by our ability to use these axioms correctly. But this only leads to a new question. In what sense could *X* vouch that *Y* uses axioms correctly, unless *X* already understands their correct use? But where does *X*'s understanding reside, and how do we know that *X* understands what he claims to understand?[28] I suspect that most materialists, like myself, would concede that *X*'s brain is the source of his 'understanding' even if this view is not conceded by idealists and interaction dualists. Thus, as far as materialists are concerned, neuropsychological theories could explain, in terms of hypothetico-deductive theories of postulated brain mechanisms, what is meant by 'understanding' all sorts of things (not just mathematics).[29] Such *meta*mathematical (etc.) explanations, as already noted, could at best be sufficient and could only be of a very general nature, since they apply to such a wide range of 'understanding'. Thus, one could get explanations of how human biological machinery could account, in terms of possible functions of that machinery, for various types of 'understanding'. Similarly one could account for a wide range of other 'rational' forms of behaviour (i.e. forms of behaviour which are provisionally, by social consensus, labelled as being 'rational') in terms of theories of brain mechanisms. How does one benefit by doing this? I believe that the major benefit, apart from possible great gains in biological insights, is to further one's confidence in materialism. The more aspects of mentality one can explain in terms of the hypothesized functioning of material components of organisms, particularly man, the more one strengthens the views of adherents of materialism. (See also my remarks in Section 14.3 below.)

Having argued above that, after all, theories about the manner in which brains transact cognitive procedures, such as doing mathematics and understanding it, *can be illuminating* (particularly for materialists), I must hasten to remove one possible misunderstanding, which I have repeatedly stressed. Theories of brain mechanisms do not explain the nature of mathematics *per se*, or for that matter the nature of anything, unless additional specific *ad hoc* hypotheses (relating to specific aspects of mathematics, say, deduction procedures etc.) are invoked. Moreover, the standards by which we judge mathematical proofs (or our criticisms of these standards) certainly do not depend on *theories* about brain mechanisms, but on people who use actual brain mechanisms. These standards, like the standards for judging all rational activities, are intrasocially variable. In order to assess the correctness of a highly sophisticated mathematical proof, this proof has to be examined by an adequately 'rational' person, i.e. somebody who has the required level of mathematical competence, as judged by appropriate consensus of opinion. In fact, most very complex mathematical proofs are usually examined by many highly proficient mathematicians, and it is their collective consensus of the validity of a proof that guarantees its ultimate (although always questionable) acceptance. Consensus can be obtained because the species specifically wired-in brains of members of the species *Homo sapiens* are in many cases sufficiently similar as regards functional potential so that when exposed to a similar culture and its problems they can reach similar conclusions. Nevertheless, explaining and understanding how human brains work, only clarifies why consensus could occur, but does not ensure either the presence of rationality in individuals or guarantee consensus on any topic.

14.3 Pragmatism versus 'critical rationality'

I have already answered implicitly the second question posed at the start of this paper. It was answered by suggesting that the criteria by which the supposed presence or absence of rationality is adjudicated are not absolute, but relative to those who regard some reasoning, action or decision (etc.) as 'acceptably meaningful and plausible'. The people who are relied on for establishing consensus may be a few experts or a large group depending on circumstances. I shall now turn to the third question posed at the beginning of the paper. It concerns the means whereby the criteria (i.e. the consensus of opinion) concerning the rationality or otherwise of something are defended, rejected or replaced.

I have already dealt with this topic elsewhere,[30] but I propose to repeat some bits of my earlier argument, and shall reach some new conclusions.

We must ask whether rationality standards, when based on partial or large-scale public consensus, by people simply agreeing that such and such reasoning or behaviour is rational, are defensible. After all totalitarian governments try to enforce public consensus by propaganda or forced labour camps (etc.). Bartley[31] has drawn attention to the views of sceptics and fideists, who try to demonstrate that 'from a rational point of view, the choice between competing beliefs and positions and ways of life, whether scientific, mathematical, moral, religious, metaphysical, political, or other, is arbitrary', and that 'rationality' then helps to make non-arbitrary choices, and this is where decision making may play a prominent role.[32] Bartley correctly stresses[33] in agreement with Popper[34] that 'The traditional questions of philosophy are authoritarian in structure in the sense *that they all beg authoritarian answers*. Questions like 'How do you know?' 'How do you justify your beliefs?' or 'With what do you guarantee your opinions?' demand authoritarian answers, whether those authorities in particular be the Bible, the leader, the social class, the nation, the fortune teller, the word of God' ... and so forth. Yet, if, as I have argued at length, consensus of opinion is at the basis of publicly acknowledged 'rationality' then such consensus can always be labelled authoritarian. Consensus is, of course, or should be, open to criticism, and this is where totalitarianism fails miserably.

Bartley[35] has advocated 'nonjustificational criticisms' as a means of combating intellectual error (or 'irrationality'). This, however, amounts once more to accepting authorities of a different type, namely the views of critics. At first sight this may not be obvious, since according to Bartley[36] nonjustificational criticisms (such as test statements of scientific theories) are intended to be *hypothetical* 'and criticisable and revisable, just like anything else in the system there is no justification, no proof, no fixed point anywhere in the system'. For practical purposes, however, *pace* Bartley, individuals or a group (such as committees or groups of scientists) often have to make decisions or by gradual or rapid agreement establish a consensus of opinion. So *de facto* it appears that one cannot get away from society-bound rationality whether constructive or critical. Indeed, criticism has sometimes turned out to have been totally irrational but very influential for long periods. It may strike some people as utopian to imagine that maximum exposure to criticism of existing consensus leads to the elimination of intellectual error. In practice criticism is often as authoritarian as social consensus. Personal prejudices or biases can introduce serious irrationality into the arguments of a critic. Many criticisms of scientific theories or discoveries turned out to be inappropriate and to have retarded the progress of science, particularly if the critics were regarded as leading authorities. The (then) new anatomical descriptions of Vesalius were the subject of derision in his day. People who attacked his findings based their criticisms on the authority of Galen's writings[37] (i.e. on an authority), although Galen apparently described at times the anatomy of pigs rather than that of man, without stating this. Again, a wave theory of light was propounded by Hooke, and, subsequently, in a superior version by Huygens.[38] Yet 'the rejection of the wave theory [of light] on the authority of Newton led to its abeyance for nearly a century' (quoted from op. cited note 38 p. XXI). The critique of Newton's followers was founded on the corpuscular theory of light, which explained certain optical phenomena in terms of the straight line motion of particles. It was not until Fresnel's work appeared[39] that a wave theory of light was accepted.[40]

While Popper (and Bartley) believe that progress in science and other human pursuits relies on conjectures and refutations[41] (or criticisms), the most important contributions of rationality consist of constructive efforts, such as the creation of new works of science and art, and criticism is not the primary creative source, but may help to eliminate some irrationality (but may also create some). The utopian approach to rationality by Bartley (op. cited note 1) might raise eyebrows of those social psychologists who have stressed the role of prejudice in human transactions.[42] There appear to exist rigid types of personalities, readily given to prejudices of all kinds. It could hardly be expected that these people would be prepared to consider all of their views as being only tentative and open to ready correction in the face of valid criticisms. On the other hand there are alternatives, notably pragmatism, which, while often rational (working by consensus), requires no need for regressive justifications of any procedures, but judges these by *results*. If, according to Popper[43] 'rationalists are those people who are ready to challenge and to criticize everything including...their own tradition' then I suspect that this type of paragon of virtue exists rarely, if at all, in human society, with its inbuilt prejudices and strong value-bound preferences.

14.4 Brains and rationality

Let me now briefly return to the issue of brains and rationality. Flew[44] and many of his contemporaries have argued validly that 'whatever is said about physiological mechanisms or lack of them, it is never those but only persons who act or choose'. This form of holism only asserts that in order to assess actions and choices we must observe organisms that, apparently, can act and choose at an appropriately complex level of their organization. Just as it is a watch and not its internal mechanism that indicates the time, and it is a motor car that moves along the road and not its engine, or its gear box, or its differential, so it is the integrated person whom we observe when we refer to actions and choices. Yet, from these considerations we must not conclude that neuropsychology cannot significantly illuminate how our biological constitution could enable most of us to be sometimes rational. One must distinguish between the 'biological roots' of rationality and the manifestations of rationality at the level of 'persons'. This distinction becomes sometimes relevant when dealing with abnormal 'persons', say, schizophrenics. Resort to 'holistic' communicative therapy is then often quite ineffective in restoring the rationality of afflicted patients. Drugs, which obviously affect brain machinery, as well as 'persons', seem more effective than communicative therapy with schizophrenics.[45] It is therefore imperative to understand what neuroleptic drugs do to brains of these 'persons'. More importantly still, we must try to discover the biological defects which underlie the irrationality of chronic schizophrenics. For this purpose, however, and more generally for the purpose of enriching our understanding of how material constituents could help to explain our capacity to be rational, we shall have to understand how brains can produce normal cognitive constructs. It is little comfort to a chronic schizophrenic to be told that he is a 'person'. The slogan that people are 'persons' refers to the holistic level, but does not provide deep insights into what makes people persons. To obtain such insights one must also try to understand, at least in principle, how the biological constitution of people enables them to think, act and decide as 'persons'.

14.5 Epilogue

A recent collection of essays on 'rationality and relativism'[46] contains several pro-relativist papers, as well as other papers which are anti-relativistic. For instance, Newton-Smith[47] argued that sentences that have equal meaning (when translated, if necessary) for different social groups must have the same truth value for all groups. This, although correct, disregards the fact that many important statements made within any one society are not, and do not claim to be, sentences made with truth values. Thus, the axioms of mathematics, and the starting hypotheses of scientific hypothetico-deductive systems are assumptions which cannot be proved to be true (but may at best, in the case of axioms, be shown to be consistent, or, in the case of scientific hypotheses, to be falsifiable). Translating the scientific hypotheses known as 'Newton's laws of motion' into the languages of all kinds of societies does not make these hypotheses provably 'true' in any of these languages, although these hypotheses remain acceptable by consensus (subject to their limitations) to those social groups of experts that 'understand' them, in the sense discussed in Section 14.2. Likewise, as I argued earlier in this paper, deductive reasoning is relative to those who 'understand' and accept it. Hence, the deductions of a scientific hypothetico-deductive theory from its unprovable starting hypotheses, while often acceptable in terms of the logic of a social group of experts, are no more true, in an absolute sense, than the starting hypotheses from which they derive. (A group of experts may often overlap with other social groupings. For instance eminent physicists are widely scattered among many nations). Newton-Smith appears to have disregarded that many belief systems (for instance, scientific theories) either do not claim to be absolutely true, or, in the case of religions, claim that their asserted 'truths', while not provable, are matters of faith. Moreover, the types of inferences people make and accept depend on the type of logic (and/or) mathematics that they use (cf. the example of the Congo native cited in Section 14.2 above). I have argued elsewhere in this paper (see note 18a) that the case for relative rationality does not have to rest on observations and interpretations of anthropologists, because there exist numerous examples of group-bound rationality within highly developed Western societies. Accordingly Sperber's[48] attack on the source material and interpretations of anthropologists (with respect to relativism of rationality), while interesting, fails to destroy the central arguments of the case for relativism.

Summary

It is argued that 'rationality' is something distinct from intelligence and goal-directed thinking. Attempts to formalize theoretical and practical rationality are criticized, and it is argued that a satisfactory formal definition of 'rationality' seems unobtainable, mainly because of the differences between relativists and anti-relativists. It is suggested that any alleged 'rationality' of people is not factual but hypothetical, and, like all hypotheses, hypothesized rationality cannot be proved to be true. Its alleged 'truth' can at best be a matter of consensus. The kinds of so-called 'rationality' present among similarly reasoning people depends on educational contingencies as well as on the constitutions of the brains of such people. Neuropsychology may help to explain how similarly constructed brains could reach consensus. It is argued that consensus among a large social group, or often a small body of experts, or a body of believers in the case of a religion, is essential if some mode of reasoning, action or decision is to be classified as 'rational' in the sense of being 'acceptably meaningful and plausible' to the group(s) of people concerned. Popper's views on critical rationality are rejected in favour of pragmatism, and Newton-Smith's anti-relativistic arguments are briefly criticized. Sperber's critique of anthropological research in relation to the relativity of rationality, although interesting, leaves the central contention of relativism intact.

Notes

1 E.g. Bartley, W.W., III The philosophy of Karl Popper, Part III; Rationality, criticism and logic, *Philosophia*, 1982, **11**, 121-221.

2 E.g. Castellan, Jr., N.J., Pisoni, D.B. and Potts, G.R. (eds), *Cognitive Theory* (vols. 1 and 2), Hilsdale New Jersey: Lawrence Erlbaum Associates, 1977.

3 Slater, E. and Roth, M. *Clinical Psychiatry* (3rd ed. revised) p. 692, London: Baillière Tindall, 1977.

4 Wassermann, G.D. Criticl Study of *Teleology* by A. Woodfield (Cambridge: Cambridge University Press, 1976), *Philosophia*, 1981, **10**, 125-132.

5 Whiteley, C.H. *Mind in Action*, An Essay in Philosophical Psychology, p. 75. Oxford: Oxford University Press, 1973.

6 Cf. Wassermann, G.D., *Brains and Reasoning*, London: Macmillan, 1974, or Wassermann, G.D., 'On the nature of the theory of evolution', *Philosophy of Science*, 1981, **48**, pp. 416-437 for detailed discussions of the structure of scientific theories and of theories of mechanisms.

7 Cf. Wassermann, G.D., *Neurobiological theory of Psychological Phenomena*, London: Macmillan, 1978, Chapter 6 for a review of theories, such as perception theory, that are based on probabilistic mechanisms of information representation.

8 Cf. Agassi, J., Critical Study of *The Secret of Carnap: Rudolf Carnap Logical Empiricist* (Hintikka, J. Ed.), Dordrecht: Reidel, 1975, *Philosophia*, 1981, **10**, pp. 57-88, see p. 82.

9 Op. cited note 5, p. 81.

10 Cf. my general assessment of psychoanalytic theory in op. cited note 6 (1974) or in Wassermann, G.D., Human behaviour and biology, *Dialectica*, 1983, **37**, pp. 169-184; see also Cioffi, F., Freid and the idea of a pseudo-science, in: Borger, R. and Cioffi, F. (eds.), *Explanation in the Behavioural Sciences*, Cambridge: Cambridge University Press, 1970, pp. 471-499 for a critical evaluation of psychoanalytic and similar frameworks.

11 Lee, T., Seeman, P., Tourtelotte, W.W., Farley, I.J. and Hornykiewicz, O., Binding of3H-neuroleptics and 3H-apomorphine in schizophrenic brains, *Nature*, 1978, **274**, pp. 897-900; Hornykiewicz, O., Brain catecholamines in schizophrenia a good case for noradrenaline, *Nature*, 1982, **299**, pp. 484-486. For opposed arguments see *Nature*, 1983, **305**, No. 5935; Tsuang, M.T., Genetic factors in schizophrenia, in *Biological Foundations of Psychiatry* (Grenfell, R.G. and Gabey, S., eds.), New York, Raven

Press, 1976; Kaplan, R.A. (ed.) *Genetic Factors in Schizophrenia*, Springfield, Ill: Charles C. Thomas, 1972; Heston L.L., The genetics of schizophrenia and schizoid disease, *Science*, 1970, **167**, pp. 149-156; Heston, L.L., Genes and schizophrenia, In *Biological Psychiatry Schizophrenia and Genetics*, New York: Academic Press, 1972.

12 As suggested in op. cited note 7.

13 Von Domarus, E., Über die Beziehung des Normalen zum Schizophrenen Denken, *Archiv für Psychiatrie*, 1925, **75**, pp. 641-646; Von Domarus, E. The specific laws of logic in schizophrenia, In: *Language and thought in Schizophrenia: Collected Papers* (Kasinin, J.S., ed.), Berkeley: University of California Press, 1944.

14 Arieti, S., *Interpretation of Schizophrenia*, New York, Basic Books, 1974.

15 Op. cited note 14, p. 232.

16 Storch, A., *The Primitive Archaeic Forms of Inner Experiences and Thought in Schizophrenics*, New York and Washington: Nervous and Mental Disease Publication Company, 1924.

17 Werner, H., *Comparative Psychology of Mental Development*, New York: International Universities Press, 1957. Skeptics could, of course, argue that Lévy-Bruhl's Congo native was a schizophrenic, since such a coincidence could not be ruled out.

18 As Arieti (op. cited note 14, p. 233) put it 'obviously a common characteristic or predicate (having an evil spirit) led to the 'identification''.

18a Sperber, D. (Apparent irrational beliefs, In op. cited note 46 below, pp. 149-180) has argued that anthropological reports concerning the modes of thinking of the 'beliefs' of certain non-Western societies have to be treated with great care, or must even be discarded. Yet, my conclusions concerning the society-bound (or group-bound) nature, and, hence, relativism, of putative 'rationality' can also be inferred from the existence of numerous groups with strongly entrenched belief systems *within* Western and other societies. For instance, the views of creationists *vis-a-vis* advocates of neo-Darwinian evolutionary theory provide a typical example of coexisting groups with totally incompatible belief systems within (often) the same society (e.g. nation). (cf. Kitcher, P., *Abusing Science*, Cambridge, Mass.: MIT Press, 1982; Newell, N.D. *Creation and Evolution*, New York: Columbia University Press, 1982; Futuyma, D.J. *Science on Trial*, New York: Pantheon, 1983; Godfrey, L.R. (ed.), *Scientists Confront Creationism*, New York: Norton, 1983; Eldredge, N., *The Monkey Business*, New York: Washington Square Press, 1982; Young, D.A. *Christianity and the Age of the Earth*, Grand Rapids, Mich.: Zondervan, 1982; Bambach, R.K.

(Response to Creationism, *Science*, 1983, **220**, pp. 851-953) after reviewing the books just cited, noted that

'Although evolution does not attack religion it poses the problem of what to do when convincing scientific conclusions come into conflict with the beliefs that people hold without evidence. The creationists are no more ready to deal with evolution than the Catholic church was prepared to deal with Copernican (heliocentric) astronomy in the time of Galileo'.

19 Cf. Hamilton, M., *Fish's Schizophrenia* (2nd ed.), Bristol: John Wright and Sons, 1974, see their Chapter 3; Op. cited note 14.

20 Thorn, G.W., Adams, R.D., Braunwald, V.E., Isselbacher, K.J., and Petersdorf, E.G. (eds.), *Harrison's Principles of Internal Medicine* (8th ed.), London: McGraw-Hill Kogakusha, 1977, p. 1966.

21 Goldstein, K., *The Organism*, New York: American Book, 1939; Goldstein, K., The significance of psychological research in schizophrenia, *Journal for Nervous and Mental Diseases*, 1939, **97**, pp. 261-279.

22 Cf. Chomsky, N., *Language and Mind*, New York: Brace Janovich, 1972; Piatelli-Palmarini, M. (ed.), *Language and Learning*: the debate between Jean Piaget and Noam Chomsky, Cambridge, Mass.: Harvard University Press, 1980; Chomsky, N., *Reflections on Language*, New York: Pantheon Books, 1975; Lyons, J., *Chomsky* (revised ed.), New York: Fontana Modern Masters, 1977.

23. Op. cited note 10 in *Dialectica*.

24 Kekes, J., Critical study of *Science in Flux* by J. Agassi, Dordrecht: Reidel, 1975, *Philosophia*, 1981, **10**, pp. 43-56, p. 45.

25 Jung, C.G. and Pauli, W., *The Interpretation of Nature and Psyche*, London: Routledge and Kegan Paul, 1952, Chapter 2.

26 See op. cited note 6, 1981, p. 423 and 1974, p. 28 and p. 37. In particular these two works try to explain what is to be understood by a 'theory of a mechanism'. See also chapter 2 above.

27 Sutherland, N.S., Basis for an ethical system, *Nature*, 1979, **279**, p. 823.

28 I have dealt elsewhere with closely related aspects of the philosophy of mind, and have formulated a novel version of epiphenomenalism (which could deal on a materialistic basis with conscious experiences that may be present in 'understanding'), see Wassermann, G.D., Biological aspects of the philosophy of mind, *Metaphilosophy*, 1980, **11**, pp. 199-209; Wassermann, G.D., Reply to Popper's attack on epiphenomenalism, *Mind*, 1979, **88**, pp. 572-575; Wassermann, G.D., Materialism and Mentality, *Review of Metaphysics*, 1982, **35**, pp. 715-729.

29 Cf. op cited note 7 for extensive attempts to formulate such theories.

30 Op. cited note 6 (1974) Section 1.9.

31 Barthley, W.W., III Rationality versus the theory of rationality, In: *The Critical Approach to Science and Philosophy* (Bunge, M., ed.), London: Collier-Macmillan, 1964. p. 3.
32 Cf. Watkins, J., Imperfect rationality, In: *Explanation in the Behavioural Sciences* (Borger, R. and Cioffi, F., eds.), Cambridge: Cambridge University Press, 1970, pp. 167-217.
33 Op. cited note 31, p. 20.
34 Popper, K.R., On the sources of knowledge and of ignorance, *Proceedings of the British Academy*, 1960, **46**, pp. 39-71.
35 Op. cited note 31, pp. 29-31 and op. cited note 1.
36 Op. cited note 1 Section 14.
37 Cf. Haggard, H.W., *The Doctor in History*, New Haven: Yale University Press, 1934.
38 Born, M., and Wolf. E., *Principles of Optics*, London: Pergamon Press, 1959, p. XX.
39 Whittaker, E., *A History of the Theories of Aether and Electricity. The Classical Theories*, Edinburgh: Nelson, 1951, p. 108.
40 For additional examples of misdirected scientific criticisms see op. cited note 6 (1974) Section 1.9.
41 Cf. Popper, K.R., *Conjectures and Refutations*, New York: Harper and Row, 1968.
42 Allport, G.W., *The Nature of Prejudice*, Cambridge, Mass.: Addison-Wesley, 1954.
43 Op. cited note 41, p. 122.
44 Flew, A., *A Rational Animal*, Oxford: Clarendon Press, 1978, Chapter 5.
45 In a well documented review P.R.A. May (Psychotherapy and ataraxic drugs, In *Handbook of Psychotherapy and Behavior Change* (Bergin, A.E. and Garfield, S.L. eds.), New York: John Wiley, 1971), after surveying a variety of controlled studies, concluded 'that in general there is little difference between psychotherapy plus drug and drug therapy alone for hospitalized psychotic patients (but not for neurotic patients)'. May also considered several studies which showed that for schizophrenic in-patients drug therapy alone was better than psychotherapy alone.
46 Hollis, M. and Lukes, S. (eds.), *Rationality and Relativism*, Oxford: Basil Blackwell, 1982.
47 Newton-Smith, W., Relativism and the possibility of Interpretation, In op. cited note 46, pp. 106-122.
48 Op. cited note 18a.

This paper appeared first in *Methodology and Science* (1984) **17**, 50-69.

15 Morality and determinism

15.1 Introduction

This paper is intended as a contribution to a recent vigorous debate in *The Times*, between the distinguished journalist Bernard Levin, the eminent Oxford economist Wilfred Beckerman and the Archbishop of York, John Habgood, among others. The debate concerns morality, 'free will' and determinism. As a former German (now British) Jew, who lost close relatives at Auschwitz (e.g. my mother's sister) and who suffered personally severely in my youth under daily virulent Nazi persecution (even from schoolmasters who earlier professed to be 'good democrats' but were vicious anti-Semites), I obviously cannot remain strictly detached and neutral. Yet, I shall attempt to retain as much neutrality as possible, since I think that the main rivals in this debate have all some very relevant, interesting and valid things to say. Let me also state other, probably very relevant, biases. I am an ardent Zionist (and have been so since the experiences of my youth). In addition, I am a die-hard mechanistic materialist as regards basic philosophy, although I am tolerant of other people's religious feelings, because I realize that my materialism is as metaphysical as their religious views. With this as background let me return to the technical issues. Obviously, in a philosophical journal one can write at a level above that of *The Times*, where there is, perhaps, insufficient room to debate philosophical, biological, physical and other niceties in some depth.

As a mechanistic materialist I have a metaphysical belief that all things in the universe are composed of matter, and that the behaviour of these things is entirely a consequence of the modes of their material composition. Even our mental experiences, such as our *qualia*, our sense of values (say 'good' or 'bad'), our sense of responsibility, and so forth are, *in my opinion*, necessary concomitants of certain central nervous transactions of appropriate brain matter.[1] According to this view our wishes and intentions (etc.) are *not* causes of our actions. These actions are assumed to result entirely from brain transactions and other physiological processes. In addition it is assumed that some of the action-triggering brain transactions also produce, *epiphenomenally*, our feeling that 'we', as persons, wished or intended these actions. (Advocates of mind/brain 'identity theories' believe in a different, but closely related, doctrine). Put differently, a mechanistic materialist can regard people as self-organizing sentient automata (and this applies also to some of their subcomponents, such as brains), which can *in principle* be understood entirely in terms of appropriate *scientific systemic models*.[2] Suitable scientific systemic models explicate, among other things, what precisely we mean by matter. Typically, such models try to deduce from appropriate hypotheses the observable properties of hypothesized material constituents, such as quarks, electrons (etc.) and the assumed forces between these constituents, such as the electro-weak interactions, the strong (gluon-mediated) interactions and gravitational interactions.[3] Other scientific systemic models try to explain the manner in which increasingly complex material systems are built up from simpler ones. Accordingly, at one extreme we have nowadays scientific systemic models that explain the properties of (hypothesized) things like electrons, quarks, gravitons, W and Z bosons (etc.) in terms of 'superstring theory'.[4] At the other extreme people, like myself, have put forward scientific systemic models for brains, which suggest, for instance, how brains in terms of their hardware could represent the physical aspects of cognitive constructs, such as *Gestalten* and concepts, and how brains could represent transactions among such constructs.[5]

Taking the preceding mechanistic materialistic views as the basis of my own attitudes towards the debate between Levin, Beckerman *et al.*, I shall now turn to Levin's first article[6] since it started the ball rolling.

15.2 The evil that some men do is born within them

Levin stressed what he believes are 'two enormous and vital truths':

> The first is that some people are evil; *are* evil, not *are made* evil. The greatest and most dangerous lie of our time is that we are *solely* the result of our upbringing, our milieu, our physical surroundings, our schooling, our degree of affluence or poverty, our employment or employment prospects, our social or familial relationships and our sex lives, along with the weather, the threat of war, other people's smoking, Sellafield and the Freemasons.

Levin's distinction between *being* evil and *being made* evil demands considerable discussion. I have discussed elsewhere the substantial and intricate evidence that human beings (here considered as an animal species) and, to a lesser extent, lower species of animals, depend on their genetics *and* on environmental factors, as far as development and behaviour are concerned.[7] For instance, nervous systems are developmentally wired-in under rigid genetic control to such an extent that, apparently, no two nerve cells of the grasshopper, and probably of man as well, are alike in their specificity, and each nerve cell (among more than ten thousand million such cells in man) is thus uniquely specified.[8] As a consequence of this, nerve cells recognize each other, during normal development, with the utmost precision, dictated by their specificities. Nerve cells therefore form the typical, highly species-specific, nervous system of each animal of those species that are endowed with a nervous system. This specificity and precision of wiring-in ensures the basic potential for a nervous system to function accurately. Nevertheless, two important points must be remembered. *First*, although man's nervous system is wired-in under rigorous genetic control, how people use their brains is not necessarily a mere consequence of the wiring-in. Unlike highly stimulus-bound lower animals, which often respond in stereotyped fashion to only a very limited set of stimulus configurations, higher animals, and man in particular, are far less stimulus-bound. This is shown by people of genius who make original discoveries, and by the highly varied forms of human behaviour. For instance, although the capacity to learn is, presumably, wired-in for each of our brains, what use we make of this capacity, i.e. how much and what we learn, depends to a large degree also on environmental factors. Some people may remember whole books or parts of books (e.g. Bernard Levin), others may remember complete music scores, and some 'memory men' remember every football team that played during the last half century. Thus, the environment *does* play an important role, notwithstanding the important role of genetic control of brain development.

What is less certain is the extent to which behaviour propensities of man are genetically wired-in. Ethologists[9] have long ago discovered and described so-called 'innate releaser reactions', which occur in response to 'sign stimuli' in many lower species. In such reactions a particular environmental stimulus configuration can 'release' a stereotyped action. For instance, a particularly coloured, and specifically located, patch on the bill of a herring gull chick's mother may release pecking behaviour in the chick. Although some 'human ethologists' have claimed that man exhibits such (presumably genetically wired-in) innate releaser reactions, only a few such instances are known. Not even human coitus is stereotyped, a point harped on by those who have written quasi-textbooks on different modes and positions in which humans can copulate.

A *second* important point is the genetic interindividual variability of human (and animal) nervous systems. It is believed that each nerve cell may be specified by a *unique combination* of dozens, or possibly hundreds, of different, genetically-determined, cell-specifying proteins (cf. op. cit. note 8). These cell-specifying proteins may help to provide a unique labelling of the nerve cell's surface membrane.[10] Detailed combinatorial estimates suggest (Clowes and Wassermann, op. cit. note 10) that the wiring-in of a human brain requires considerable numbers of genes, possibly ten to twenty thousand. The important point is that any, possibly most, of these genes may exist in the population in several variant editions per gene, although some of these genes might only occur in single editions per population, because of strong eliminative natural selection pressure against carriers of other variants of such genes. Not too much is known about the precise ratios of multivariant-edition genes to single-edition genes per population for most genes of most populations. 'Selective neutralists', like Kimura,[11] believe that there may be more genes with viable variant editions in populations than other population geneticists allow for. Just the same, there can be no doubt that within a population plenty of genetic variation among the cell-specifying proteins of brains occurs, resulting in considerable interindividual differences of the wired-in aspects of brains, quite apart from interindividual variations of brains produced by individual learning.

That the normal functioning of brains depends on the presence of specific 'normal' variants of genes, that can occur in variant editions, is suggested by findings relating to mental illnesses such as schizophrenia[12] (which seems to have a demonstrable genetic basis. I wrote in this context (op. cit. note 12):

> That chronic schizophrenia...involves a genetic abnormality (as well as non-genetic precipitating factors[13]) is suggested (a) by concordance rates [i.e. coincidence rates] of chronic schizophrenia for pairs of monozygotic twins [i.e. so-called 'identical twins'], reared apart, with those for comparable dizygotic twins [i.e. non-identical twins]; and (b) observations of relevant mental abnormalities in close relatives of chronic schizophrenic patients.[14] Some cases of *acute* schizophrenia, however, show no family incidence.[15]

Thus, even in an apparently strongly genetically predisposed illness, like schizophrenia, the concordance rate for the illness for 'identical twins' is not one hundred per cent. Hence, people who carry the same 'faulty' variants of genes (for schizophrenia) need not necessarily all develop the illness, and additional environmental precipitating factors (e.g. possible stress, or possible virus infection) are required (cf. Pollin, op. cit. note 13).

Similar remarks could apply to other forms of mental disorders that are believed to have, or that could have, a heritable factor. Of particular concern, in the present context, are psychopathic states of people diagnosed as 'psychopaths'.[16] Many typical psychopaths are amoral and/or asocial personalities who are either predominantly aggressive, or predominantly inadequate or passive, or predominantly creative, although some of these character traits may coexist in various psychopaths (cf. op. cit. note 16 pp. 395ff). To cite Henderson and Gillespie (op. cit. note 16 pp. 392-393):

> We thus have people in our midst, semi-insane or semi-responsible, if you like to use such terms, who [may exist] in every walk of life from the highest social and political offices to the ranks of the...delinquent who not only are unable to conform adequately in their own personal lives, but may even be responsible for some of the greatest social crimes in history.

Psychopaths must not be confused with schizophrenics, although some schizophrenics may, in addition to, or as part of, their illness, exhibit quasi-psychopathic traits. To what extent prominent psychopaths (e.g. Adolf Hitler and Josef Stalin) had a genetic factor in their make-up remains unknown. In fact, for most psychopaths genetic patterns, related to their psychopathic states, either remain unknown, and perhaps non-existent, or are certainly not as well established as the genetics of schizophrenia. Hence, to be as certain as Bernard Levin and to assert that some people *are* evil (presumably by heredity) and are not *made* evil (by the environment) remains problematic. In fact, some eminent sceptics with strong left-wing leanings, notably Lewontin and Rose, have tried very hard to demonstrate, by far-fetched arguments, that even the twin studies (etc.) for schizophrenia are not reliable, and that chronic schizophrenia may not be a genetically based disorder.[17]

If people are as strongly ideologically committed to views of the left as, apparently, Lewontin and Rose, then it is not surprising that such people try to explain away, by all means fair or foul, every morsel of evidence that suggests that some of our mental capacities and some of our mental illnesses are genetically influenced or predisposed. This is where Bernard Levin's view about the 'greatest and most dangerous lie of our time' is borne out in practical terms by the desperado efforts of many neo-Marxists, who are totally sold to the idea that we are completely shaped by our environmental stimuli, so that genetic predispositions for actions and behaviour can be, and must be, completely ruled out.

Levin (op. cit. note 6) also observes correctly that:

> In any urban riot in this country, those doing the rioting are a small minority of those who live in the area of the riot; the peaceable majority, however, live in the same conditions as the violent few, and are therefore subject to the same riot-inducing pressures. Why do they not join in?

Levin may be right in suggesting that the 'few' rioters who are ringleaders are psychopaths (although he does not put it in these words). Yet, this does not establish whether their psychopathic states have a partially genetic basis, or whether these people have been entirely conditioned by learning to behave as brutally as they do. (The latter alternative might be adopted by followers of Skinner's earlier extremism.) Levin's observations could still be compatible with the conditioning hypothesis, unless it can be shown that typical ringleaders of riots come from families with long family histories of psychopaths, or if there are other strong pointers towards a genetic basis of their psychopathic states. I have not come across reliable reports on this topic. Nevertheless, if it could be shown by future research that certain kinds of (non-schizophrenic) psychopaths have a predisposing genetics, and if relevant variant genes for their disorder could be detected, this could be very useful. It should then, in principle, be possible to establish tests for screening the population for such genes, and, perhaps, even cause early abortions of foetuses at risk. Screening procedures could then conceivably provide an important safeguard against future monsters, like Hitler and Stalin and mass-murderers, pursuing their deadly impact on our planet, and prevent some of these psychopaths from enslaving masses of people and causing much misery. Without doing any harm to such psychopaths, one could then, by knowing their genetic predisposition, keep a safe eye on them and prevent them, as far as possible, from doing harm to mankind.

Although, for all we know, many psychopaths may be genetically strongly predisposed, only further research can clarify this issue. Currently it remains unknown whether most or even much psychopathic behaviour has a genetic basis. What is undeniable is the *persistence* of psychopathic traits in some of those afflicted. For instance, Adolf Hitler thought of the 'Final solution' (i.e. the extermination of as many Jews as possible) as far back as 1922, about twenty years before it was implemented. Gerald Fleming[18] notes that:

> When Josef Hell [a retired major and journalist] asked Hitler in 1922 'What do you intend to do if you should attain full freedom of action against the Jews?' Hitler, who until then had remained calm and restrained answered in a totally different manner. He did not look at me but above me into the void and made his further declarations with increasing vocal exertion, so that he got into a kind of paroxysm and finally shouted at me, as if I were a complete popular gathering: 'When I really gain power then the destruction of Jews will be my most important and first task. As soon as I have the power to do so I shall, for instance, erect gallow beside gallow in the Marienplatz in Munich, as many as the traffic allows to erect. Then Jews will be hanged, one like the next, and they will remain hanged until they stink. They will remain hanged as long as it is possible according to the principles of hygiene.

> As soon as one has removed them from the gallows it will be the turn of the next lot, until the last Jew in Munich has been exterminated. Similarly one will proceed in other towns until Germany has been rid of the last Jew.'

This is closely similar to what happened, if by different technical means, in the extermination camps of Auschwitz, Treblinka (etc.) during World War II, at the instigation of the most murderous psychopath the world has ever seen.

What percentage of the currently living members of the species *Homo sapiens* are actual or potential psychopaths remains unknown, although the frequently reported murders, committed by single individuals (as well as other serious crimes) could provide some indicators. Yet, it remains true that, because of modern, beastly, methods of mass-extermination, a relatively small number of individuals were sufficient to murder millions of Jews in the German death camps during World War II. It has been argued by Hannah Arendt and Bruno Bettelheim[19] that there was a banality about organizers of mass-murder, like the German Adolf Eichmann, who carefully arranged the deportation of European Jews to extermination camps and seemed delighted with his organizational successes. Bettelheim (who himself was a concentration camp victim of the German Third Reich) claimed, rather innocently, that (op. cit. note 19, p. 259)

> For example by all 'scientific' standards Eichmann was a 'normal' person. Half a dozen psychiatrists had certified him as 'normal' 'more normal, at any rate, than I am after having examined him', one of them was said to have exclaimed, while another found that his whole psychological outlook, his attitude towards his wife and children, mother and father, brothers, sisters and friends, was not only normal but most desirable....

What these psychiatrists, obviously, did not grasp in either Eichmann's or Hitler's case, and possibly in the cases of many other psychopaths, is the fact that many a psychopath may hold 'normal' views on a very wide variety of topics, including family relationships, and yet be a monstrous psychopath. In other words, a psychopath's deviation from the norm may be very restricted (e.g. some psychopaths have only murdered prostitutes, others only Jews). The fault here, too obviously, lies with the inadequacies of currently used psychiatric and psychological methods for assessing psychopathic traits. Probably many mass-murderers would appear as 'perfectly normal' in many respects to many psychiatrists. Indeed, diagnostic criteria for this kind of personality disorder are notoriously unreliable. (Psychiatrists fail even miserably in the diagnosis of, apparently, quite straightforward cases of major mental illnesses. For instance, one consultant psychiatrist, who saw my wife after she had suffered from almost continuous endogenous unipolar depression for eighteen months, pronounced her 'perfectly normal'. Yet, within six more months she committed suicide after suffering further spells of depression.)

I conclude at this stage that, although many psychopathic personalities *could* have a genetic basis for their personality defect, no reliable information exists (except for certain schizophrenics) concerning any possible genetics of such disorders, and such disorders are also difficult to diagnose. I agree with Bernard Levin that, often, only a *handful* of psychopaths can start, and keep going, massive onslaughts on people's lives and/or their property, and that this can happen in concentration camps, labour camps, in rioting crowds or among football hooligans.

Bernard Levin (op. cited note 6) cited the following passage from Someth May's book *Cambodian Witness*[20] (a passage which reminds one of Hitler's 1922 outburst cited above):

> Comrade Tek...had worked himself into a rage. He approached the pile of struggling monkeys [he had broken their arms and tied their feet together], and killed them one by one...with a blow to the back of their skull. 'And now', he said, 'I'll show you the way I used to kill the Lon Nol soldiers when we caught them, and the way to get the liver out.' He laid the last monkey flat on the ground... . He made a cut to the stomach. Then he pressed hard on the incision with both his hands. The monkey screeched. The liver came out whole. Comrade Tek then slit the animal's throat. He said, 'If it had been a man, I would have put my foot in the cut to get the right pressure. Otherwise the liver never comes out properly.'

After this horrific account, Bernard Levin argued towards the end of his article:

> We do have a choice every minute of every day. I said that the inability of most people to behave like Tek bears witness to two great truths. I have dealt with one of them the fact that some people are evil and don't mind a bit; but the other is more important. It is this: the inability to do great evil comes as much from our own will as does the propensity to steal, assault and burn. If we do not become Comrade Tek it is because we decide not to, just as we stab a policeman because we feel like it.

It is here that Levin's *mentalism* (i.e. resort to 'the will' and to 'feeling like it' as quasi-causal agents) comes in. As a mechanistic materialist I simply reject such mentalistic talk as redundant (although perhaps useful in everyday discourse). Our decisions, our feelings and our 'willing' to do something are, according to my brand of mechanistic materialism, *concomitants* of brain transactions and not *causes* of any material transactions. A materialist could argue that if we do not all become like Comrade Tek, or Adolf Hitler, or Eichmann, or Josef Stalin and a host of others that come to mind, then it is not because we 'decide' to be otherwise, or 'will' to be 'normal', but because our brains and other somatic equipment make us function like 'normal' human beings. Words like 'will' and 'decision' are inherent in our learned everyday language. Yet, these ancestrally descended modes of speaking *do not force us to believe* that there are 'mental' (i.e. non-material) entities within us, which are something apart from brains and their autonomous mechanisms, and that it is these mental entities that make us 'will' things and 'decide'.[21]

Towards the end of his first article, Levin (op. cit. note 6) believes that there remains 'the most tremendous question of all to be asked though not answered. It is: Why do most people given the choice of being good or bad choose to be good?' A mechanistic materialist, however, does not have to accept Levin's premises. For the materialist 'choices', like 'decisions' and 'willing' are at best generic terms, derived from ancestral mentalistic habits of talking. These terms may, nowadays, be regarded, by materialists, as shorthand for certain mechanistic brain processes which (epiphenomenally) may be accompanied by our 'feelings' that we are 'making decisions', that 'we are choosing' and that our 'will' is causing us to do something, when, in reality, our brains are causing us to act in certain ways. Whether these brains are pathologically predisposed (by wrong genetics and/or wrong learning) is a different matter, but it does not imply an 'autonomous mental agent' apart from our brain. Again, for all we know, brains could provide molecular biological representations of *moral 'values'*, such as 'good', or 'bad' and much else.[22] Moreover, Bernard Levin could well be right. It could be that most brains will normally select actions whose brain representations are associated with brain representations of the value 'good' rather than actions whose brain representations are associated with brain representations of the value 'bad'. Nevertheless, this conjecture is far from obvious since the attribution of values may depend on many contingencies. A general may praise his troops for wiping out an enemy division in combat, whereas a judge may call a mass-murderer a wicked man.

Levin's question is certainly very important and can, at least in principle, be dealt with by current brain-model theorists, although the discovery of brain representations of 'moral values' at the observational and/or experimental level is probably still a long way off. I conclude that Levin has raised many important points, although his mentalistic interpretations, and many other aspects of his arguments, could be replaced by some of the alternative interpretations that I discussed.

15.3 Motivations of the anti-inheritance lobby

Perhaps the main reason why some of my fellow Jews, like Lewontin and Rose (op. cit. note 17), but not Bernard Levin, try to deny, wherever possible, any apparent *genetic* component of human behaviour, is the misplaced idea that admission of a genetic basis of, say, psychopathology, opens the door to *racism*. Did not that arch-criminal Hitler, and many millions of his German followers, proclaim that the Jews were an inferior race because they had inferior variants (alleles) of various kinds of genes? Indeed, those numerous benighted German teachers, who taught *'Rassenkunde'* (racial studies) in German grammar schools, taught precisely that kind of nonsense. To counter this, many, equally misguided, Jews strenuously denied that the Jews are a race, in the sense of forming a *specific population* (both within the state of Israel and in the Diaspora). Yet, on population-genetic grounds, it seems as likely that the Jews form an identifiable population as the non-Jewish part of the British population or the non-Jewish part of the French population, or many other populations that remain inbreeding either because of tradition, language or religion. Thus most Britishers marry other Britishers, most French people marry other French people, which may have something to do with shared language and customs. Likewise, many Jews, notably religious ones, marry Jews, because the Jewish religion *demands* that Jews do not marry out. People, like myself, who are Zionists, do not deny that Jews, because of their inbreeding, form a specific subpopulation, in the population-genetic sense, much as the non-Jewish members of the British and French peoples form subpopulations of the human species. Yet, Jews, like myself, do not accept the views held by many Germans, in Hitler's time and before it, that the genetics of Jews is inferior to that of other numerous inbreeding populations on earth such as the non-Jewish part of the German population. Nor is the Jewish genetic make-up demonstrably superior (granted, even, that the Jews have, like other populations, produced many eminent people).

I conclude that one can accept that there are likely to exist genetic specificities for different human populations on earth, without accepting racial superiorities of any one population *vis-à-vis* any other. One also does not have to accept the strenuous denials by Lewontin, Rose and others that genes play a part in shaping our personalities, although one must admit that our environment plays a strong part in human character formation. Yet, the Lewontin, Rose *et al.* view, which amounts almost to the old German saying *'Der Mensch ist was er isst'*, seems to me quite wrong-headed.

15.4 Does determinism lead to condonation of guilt?

Beckerman[23] refers to Levin's views (op. cit. note 6), already mentioned, that 'the greatest and most dangerous lie of our time is that we are solely the result of our upbringing, our milieu...', etc. Levin stressed that this is a dangerous lie, since, in his opinion, it leads to 'determinism...and the...condonation of guilt and the dismissal of responsibility'. Beckerman states that this outburst by Levin:

'is a total misrepresentation of determinism, which does not require that people's behaviour is determined solely by their environment. Their genes play a part. We would be justified in passing moral judgements only if we thought that people are somehow morally responsible for their genetic endowment.'

Although I agree wholeheartedly with Beckerman that both environment *and* genetics determine our behaviour (cf. my long discussion of this issue in op. cit. note 7), 1 disagree with some of Beckerman's conclusions, particularly his statement that:

'With increasing scientific understanding of the genetic basis of personality, the absurdity of passing moral judgements on people for acts we abhor but which we know to be the result of, say, some specific abnormality in their chromosomes, will become more obvious (op. cit. note 23).'

To start with, as I argued elsewhere (op. cit. note 7), even in cases where behaviour has, or seems to have, a distinct genetic component, we cannot say how *much* of that behaviour (or misbehaviour, as appropriate) is genetically caused and how much environmentally. Moreover, when we pass a moral judgement, say, on somebody like the Yorkshire Ripper, Peter Sutcliffe, who was a mass-murderer, but allegedly a paranoid schizophrenic (a disorder with a genetic basis), then this judgement is based on our acceptance of the canons of morality that prevail in our society. This acceptance of canons of morality is not ruled out by any knowledge we may have about behaviour and its possible partial genetic basis. We can *classify* a man as evil by virtue of what he does and irrespective of any knowledge of his genetics. Knowledge of his genetics might be relevant to the *type* of judgement given by a judge. For instance, in a conceivable future case of a known hereditarily disposed psychopath with a long record of serious crime, it may be important to lock him up permanently so as to protect society rather than to 'punish' him for his crimes.

We must also realize that although a mechanistic materialist, like myself, could argue (as part of his metaphysics) that all material events are 'determined', at least statistically in the quantum mechanical sense, this does not prevent us from forming moral judgements. In fact, according to mechanistic materialists, the forming of moral judgements is carried out by brains and is, therefore, a deterministic procedure, as much 'determined' as an act of murder. So let me discuss in some more detail in which sense, and to what extent, events in the universe are statistically determined.

In the Introduction I noted that, according to mechanistic materialists, human beings and other organisms can be regarded as automata, which can, in principle, be interpreted in terms of scientific systemic models, and that similar remarks apply to every kind of materially composed system. Scientific systemic models, however, idealize by making simplifying assumptions about the composition and functioning of 'real' systems, so that scientific systemic models do not claim to represent 'real' systems but only man-invented *models* of such systems. Within powerful hypothetico-deductive theories, such as classical mechanics, quantum mechanics and electrodynamics, people have invented myriads of scientific systemic models in terms of which they discussed, say, properties of atoms, or quarks[24] (etc.) and properties of all kinds of other complex systems. Let us look briefly, and only superficially, at just one kind of *idealization* which is typical of many scientific systemic models. Many of these models, in physics and in theoretical biology (etc.), assume so-called *'initial conditions'*. These initial conditions make *assumptions* about the 'state' of a system at a specified initial instant, say time $t = 0$. In practice 'real systems' are not isolated, and always interact with some, or possibly many, other systems most or all of the time. Hence, we cannot really determine their precise initial conditions at any specific time, but can only idealize by assuming *ad hoc* initial conditions. Accordingly, even if the scientific systemic model used is, apparently, of a non-statistical deterministic kind, then its *idealized* initial conditions assure, at best, only an idealized *non-*statistical determinism.

In many branches of science the scientific systemic modelling involves an idealized statistical determination. For instance, in quantum mechanics the 'equations of state' of systems provide scientific systemic models which only yield a statistical determination. (Thus, *given* the initial *probability* that the modelled system is at time $t = 0$ in a particular, theoretically specified, state, the equation of state of the quantum mechanical scientific systemic model used permits us to determine the probability that at any future time the system will be in another permissible specifiable state. This, in principle, provides a *modelled* statistical determinism.)

The trouble with this statistical determinism is that we can never know the precise initial state of a system, so that the *specified* initial state of a scientific systemic model is an artificial idealization, an *ad hoc assumption* and an *extra-systemic hypothesis*. This state of affairs provides a mechanistic agnosticism, i.e. a set of irremovable gaps in our complete explanation of nature. This, of course, does not mean that we have to invoke deities to fill the gaps. It only shows that our faith in determinism is as metaphysical as any belief in mechanistic materialism. Yet, I believe that faith in determinism is considerably endorsed by the excellent agreement of myriads of scientific systemic models with experiments and observations, despite the *ad hoc* initial conditions of these scientific systemic models. (It seems that certain important explanatory aspects of these models are not very sensitively dependent on the precise initial conditions that one chooses for the model, e.g. the precise instant of start of a sequence of modelled events.) This agreement of a host of familiar scientific systemic models with a (metaphysical) *belief* in determinism, encourages the view that scientific systemic (deterministic) models of how brains produce behaviour could also *explain* our brain *capacity* to make moral judgements and to exhibit responsibility.

I conclude that Beckerman's main argument, cited above, is unsatisfactory and that *any metaphysical belief in 'determinism' does not force us to consider the passing of moral judgements 'an absurdity'*. Beckerman (op. cit. note 23) also asks:

> Suppose we took two people with identical genetic endowments and identical environmental experience, would we expect them to behave differently in any particular situation?

This, alas, seems to me a very unrealistic question. To start with, for reasons stated, we could not ascertain whether the initial conditions, say at birth, of these monozygotic twins (i.e. so-called 'identical twins') were the same. Worse, for Beckerman, having 'identical genetic endowments' (technically known as 'identical genotypes') does not imply having identical 'phenotypes', the phenotypes being the detectable features. Here, I suspect, Lewontin would come down on Beckerman with a big sledgehammer. Lewontin has noted[25] that one of the biggest gaps in population genetics is the absence of a reliable relation between phenotype and genotype. Both phenotype and genotype are huge multiparameter systems, and to map one on the other has not proved possible in practice for a single phenotype and genotype of any one individual of any one species, let alone millions of species that are known to exist.

Returning to Beckerman's question, about people with identical genetic endowments and identical environmental experience, it is interesting that Beckerman complains that when he challenges philosophers, who shrink from the determinist conclusion, to give him a straight answer to his question, he is 'usually met by a smokescreen of technical sophistry'. Yet, I sympathize with these philosophers, since I believe that Beckerman has neither appreciated some of the complexities of quantum mechanics (and its kind of statistical determinism) nor, apparently, recognized the complexity of the genotype-phenotype problematics in biology. Finally, let me stress again that our *capacity to* pass moral judgements seems to me to be as much wired into our nervous systems (under genetic control) as our capacity for enacting certain behavioural aspects of sexual reproduction. A wired-in capacity, of course, does not determine what use we make of it, since that use depends also on environmental contingencies, namely what sorts of environments we encounter. Even if there were *complete* determinism (in the now defunct sense of Laplace[26]) as distinct from statistical determinism, then it would remain still feasible that parts of our brains are determined so as to make moral judgements in response to environmental events, and to cause corresponding behaviour (and epiphenomenal feelings, e.g. of revulsion, etc.). I agree, however, with Beckerman that our sense of 'freedom of will', is an illusion, in the sense that it is a brain-generated feeling of freedom. In my view, this feeling, that we are 'free agents', is as brain-created as our feeling of *reality*[27] which Michotte has experimentally extensively investigated (cf. Vernon, op. cit. note 27 for detailed accounts of Michotte's work). Likewise the capacity for a *feeling* of 'moral responsibility' could be wired into our brains, and feeling guilty does not necessarily imply that we *are* in an absolute sense 'guilty' of responsibility for, say, misbehaviour. In this respect I totally agree with Beckerman. On the other hand, the *capacity* for demanding punishment for misdeeds may also be wired into our brains and could be utilized when environmental circumstances prompt application of this capacity. More generally the *capacity* for acting morally, as well as immorally, may be wired-in through our brains and form part of the deterministic scheme. Yet, we can never show the completeness of that deterministic scheme, because of the limits inherent in our forms of theorizing (see above).

I conclude that there are no valid arguments within mechanistic materialism for denying the existence and application of moral judgements, and that Beckerman's position seems to me in parts to be as exaggerated as some parts of Levin's position or that of Lewontin *et al.* The main difficulty of Beckerman's position arises, I believe, from a failure to distinguish between two issues. (1) People are, obviously, not responsible for the ways their brains are (jointly genetically and environmentally) wired in. With this point Beckerman agrees. (2) There exists a wired-in capacity of the brains of most or all people that enables these people to *claim* that they are (or feel) responsible for their actions. Beckerman seems to have disregarded this second issue. What people claim does not necessarily correspond to realities, but could be an illusion. Thus, our sense of responsibility could, like visual illusions, arise as a result of purely material processes that make us the way we are, so that felt responsibility could feel very real but be an illusion just the same.

15.5 Human beings could be automata with 'free will'

In an onslaught on Beckerman's somewhat extreme position, Bernard Levin[28] complains that Beckerman's views imply that 'we have no free will at all, we are nothing but automata, and that the accident of our genes determines how we behave from birth to death'. I am not at all sure that Beckerman would wish to defend such an extreme position, since he would undoubtedly allow for a reciprocal interaction between genotype and environment. Moreover, the arguments of the preceding sections indicate that human beings could, perfectly well, be sentient self-organizing automata whose brains are developmentally wired in so as to have the *capacity* to think that they have 'free will' (and the wiring-in is under genetic control, which still does not decide the phenotype completely!). The outcome of this capacity may be deceptive, and an illusion. Simply because people like Levin think that we have 'free will' we cannot be sure that this alleged 'free will' is other than a make-belief. In fact, one might ask 'by what criteria would Levin's *mentalism* (see section 15.2 above) guarantee that 'free will' is something apart, some purely mental entity, which is not a product of matter?' If Levin came clean and admitted that, like Chomsky and others, he is a 'mentalist' with a bias against materialism, then his metaphysical position would be clear from the outset. Unfortunately Levin does not seem to realize that his own position is as metaphysical as that of the most strident mechanistic materialists (including people like myself), i.e. people who must not be confused with dialectical materialists. (In fact, *mechanistic* materialism is not linked to any political doctrine on the left or right, and I am politically a Tory voter.)

It does not help Levin's case to suggest, as he does (op. cit. note 28), that Beckerman should 'take three months off and spend the time reading the entire works of Professor Sir Karl Popper'. In fact, as is well known, Popper is heavily biased in this whole debate and is an 'interaction dualist' in the Cartesian tradition.[29] Not surprisingly, Popper made a vain attempt to destroy the core of one popular prop of the mental aspects of mechanistic materialism, namely epiphenomenalism. I argued in *Mind*[30] that Popper's supposed refutation of epiphenomenalism is unsatisfactory. It is, therefore, not helping Levin's cause to fall back on Popper as an ally. Indeed, I fear that some of Levin's retort to Beckerman (op. cit. note 28) amounts to personal abuse, unworthy of a highly distinguished journalist with whose views I often agree.

Levin (op. cit. note 28) insists that:

> One of the reasons that Mrs Thatcher causes such indignation is that her entire approach to society is based on her instinctive rejection of the argument that we are robots, and therefore cannot be blamed for what we do.

Yet, a sensible mechanistic materialist need not adopt as crude a view as Levin wishes to attribute to Mrs Thatcher. People are certainly not *non*-self-organizing, *in*sentient automata in the sense in which motor cars, or television sets, or industrial robots, are machines. It can be argued, but not proved, that human beings are immensely complex *self-organizing, sentient automata,* composed of very intricate self-organizing biological hardware which can function so as to *appear responsible* to other automata of the same kind (namely other people). There are no reasons why human beings, regarded as automata, should not blame each other, if they fail to conform to accepted norms of responsibility. These automata can produce brain representations of cognitive constructs and of actions and can *classify* many such actions either as responsible or irresponsible. The great thinker and economist Hayek[31] already regarded human brains as *(inter alia)* classifying automata, and I believe that Hayek, a Nobel Laureate, had a firmer grasp of neuropsychology than either Bernard Levin or Mrs Thatcher, much as I admire both of them.

Scientific systemic models for brains can try to model, in terms of assumed appropriate biological hardware, the functions, and production of psychological output, of actual brains (cf. op. cit. note 5). Yet, although these scientific systemic models for brains cannot be proved to represent truths or absolute realities, and, in fact, may not be very good models, the truths of the opposed metaphysics of Bernard Levin and (supposedly) that of Mrs Thatcher are equally unprovable. Levin (op. cit. note 28) believes that each human being has a *soul* which can point out the difference between right and wrong. This, of course, is the view upheld, directly or indirectly, by all convinced interaction dualists, including Popper and Eccles (op. cit. note 29), the Pope, the Archbishop of York, the Chief Rabbi of the United Kingdom and countless other adherents of advanced or primitive religions. By contrast mechanistic materialists regard 'the (immaterial) soul' or other metaphorical 'agents' (cf. my discussion of metaphorical agents in op. cit. note 21) as redundant pieces of metaphysics, despite the age-old historical roots of the concept 'soul'. Levin's opinion (op. cit. note 28) that:

> there is nothing so dreadful - drug-pushing, violence, rape, Aids - that it cannot be ascribed by the New Determinists, to something other than wickedness - usually Mrs Thatcher or her policies...

is just part of a dogmatic metaphysics. This metaphysics denies that there could be *sentient,* self-organizing automata, made entirely of biological hardware, and devoid of metaphorical 'agents', such as 'the immaterial soul' or Driesch's 'entelechies', and this metaphysics also denies that such automata *could be* capable of wickedness. According to Levin's metaphysics, only systems with a body *and* a 'soul' can be wicked.

Finally, some words about the Archbishop of York's views[32] on some of the preceding issues, although I believe that most of his relevant points have already been dealt with implicitly in the preceding sections. Dr Habgood finds it odd that some of the distinguished correspondents of *The Times*

> should be so wedded to determinism in that most difficult of all fields, the study of human behaviour, at a time when in the field of physics it looks decidedly shaky.

I noted earlier that quantum mechanical determinism is of a *statistical* kind, and, as such, there has been nothing shaky about this kind of determinism. Besides, I stressed that scientific systemic models, *per se*, are no guarantors of determinism. Even the best scientific systemic models are, after all, only highly idealized models. In such subjects as quantum chemistry people have tackled the structure of moderately complex molecules, but have not, without empirical aid, applied quantum mechanics to any sizeable macromolecule. Even statistical mechanics is no adequate tool for modelling *heterogeneous* systems, such as cells of organisms, let alone a complete organism. Thus, while all our known scientific systemic models are not guarantors of determinism, they are, nevertheless, at least *consistent* with it. I conclude that quantum mechanics certainly does not disprove determinism, and remains consistent with it, although belief in the *universal* operation of statistical determinism is a metaphysical doctrine. The Archbishop is certainly right that orthodox science does not claim that 'all events can in principle be predicted'. Since we do not and cannot know *in practice* the complete initial conditions of any, even moderately complex, events, practice makes nonsense of any claim of what is possible 'in *principle*'. I conclude that human morality and so-called 'freedom of will' are *compatible* with determinism in the very restricted sense in which we can talk about 'determinism' in the sciences.

15.6 Postscript

Miss Marjorie Worswick has drawn my attention to Bernard Levin's book *Taking Sides* (London: Pan Books, 1980). This book should leave no doubt that Levin is a mentalist and anti-materialist. For instance, he writes (op. cit. p. 124) that:

> Apart from a few of the sillier scientists and the more bone-headed survivals of nineteenth-century anti-clericalism, almost everybody today reacts to an assertion that the materialist explanation of the universe is obviously inadequate, and indeed ridiculous, in one of two ways. Either they assent to a proposition too obviously true to be worth discussing except as a starting point for a discussion of what is worth talking about, or they display the kind

of panic reaction that I predicted for Brian Inglis's *Natural and Supernatural,* and that, when the book was published, fulfilled my prediction ninety times over and then nine more. (The book is a history of the paranormal from the earliest times to the First World War. I predicted a terrified rejection of the evidence by those unable to face the facts that there are things in the Universe that they are unable to understand.)

Here, as elsewhere, Levin's argument, apart from some mudslinging, is based on a sad misrepresentation and misunderstanding of the powers of a thoroughgoing *modern* mechanistic materialism (cf. op. cit. notes 1, 21, 30). Moreover, there are no obvious reasons why some, or perhaps many, of the phenomena of parapsychology must necessarily be incompatible with mechanistic materialism. Even if in addition to already familiar kinds of matter, physicists, following recent speculative suggestions,[33] should accept the postulate of 'shadow matter', then, for all we know, shadow matter (say via gravitons) might play a part in explaining certain aspects of paranormal phenomena.[34] If so, the combined world of matter and shadow matter would still form a *material* universe; and materialism might be beyond its infancy but may not yet have reached maturity.

Notes

1 Wassermann, G.D., 'Materialism and Mentality', *Review of Metaphysics,* **35** (1982), 715-729.
2 Wassermann, G.D., 'Theories, Systemic-Models (SYMOs), Laws and Facts in Evolutionary and other Sciences', *Synthese,* **79** (1989), 489-514.
3 Davies, P.C.W., *The Forces of Nature,* (Cambridge University Press, 1979).
4 Green, M.B., 'Unification of Forces and Particles in Superstring Theories', *Nature,* London, **314** (1985), 409-414.
5 Wassermann, G.D., *Neurobiological Theory of Psychological Phenomena,* (London: Macmillan, 1978).
6 Levin, B., 'The Evil Some Men Do is Born Within Them', *The Times,* London, (24 November 1986), 16.
7 Wassermann, G.D., 'Human Behaviour and Biology', *Dialectica* **37** (1983), 169-184.
8 Cf. the extensive review of genetically controlled species-specific wiring-in of nervous systems by G.D. Wassermann, 'The wiring-in of Neural Nets Revisited', *Bulletin of Mathematical Biology* **48** (1986), 661-680.
9 Hinde, R.A., *Animal Behavior: a Synthesis of Ethology and Comparative Psychology,* 2nd edn. (New York: McGraw-Hill, 1970).

10 Clowes, J.S., and Wassermann, G.D., 'Genetic Control Theory of Developmental Events', *Bulletin of Mathematical Biology* **46** (1984), 785-825 and Wassermann, G.D., *Molecular Control of Cell Differentiation and Morphogenesis* (New York: Marcel Dekker 1972)
11 Kimura, M., 'The Neutral Theory of Molecular Evolution', *Scientific American* **241** (November 1979), 94-104.
12 Wassermann, G.D., 'Note on an Abnormality of the Operator for the Structural Gene of the Dopamine D1 Receptor as a Possible Partial Cause of Schizophrenia', *Journal of Theoretical Biology* **120** (1986), 277-283.
13 Pollin, W., 'The Pathogenesis of Schizophrenia', *Archives of General Psychiatry* **27** (1972), 29-37.
14 Gottesman, I.I. and Shields, J. *Schizophrenia and Genetics* (New York: Acadmic Press, 1972); Gottesman, I.I. and Shields, J., 'A Critical Review of Recent Adoption, Twin and Family Studies of Schizophrenia: Behavioral Genetics Perspectives', *Schizophrenia Bulletin* **2** (1976), 360-401; Heston, L.L., 'The Genetics of Schizophrenia and Schizoid Disease', *Science* **167** (1970), 249-256; Heston, L.L., in *Biological Psychiatry*, J. Mendels (ed.) (New York: Wiley-Interscience, 1973); Crowe, R.R., in *Schizophrenia as a Brain Disease*, Henn, F.A., and Nasrallah, H. (eds.) (New York: Oxford University Press, 1982), 40-60.
15 Berrettini, W., Vogel, W.H. and Clouse, R., 'Platelet Monoamine Oxidase in Chronic Schizophrenia', *American Journal of Psychiatry* **134** (1977), 805-807, and Crowe, R.R. (1982) op. cit. note 14, pp. 45-46.
16 Henderson, D., and Gillespie, R.D., *A Textbook of Psychiatry*, 7th edn (London: Oxford University Press, 1952).
17 Lewontin, R.C., Rose, S. and Kamin, L.J., *Not in Our Genes* (New York: Pantheon, 1984).
18 Fleming, G., *Hitler und die Entlösung* (München: Limes Verlag), 29 (English translation *Hitler and the Final Solution* (London: Hamish Hamilton, 1985); the present translation is provided by the present author).
19 Arendt, H., *Eichmann in Jerusalem: A Report on the Banality of Evil* (New York: The Viking Press, 1963); Bettelheim, B., *Surviving and other Essays* (London: Thames and Hudson), 258-273.
20 May, S., *Cambodian Witness: the Autobiography of Someth May* (London: Faber, 1986).
21 See also Wassermann, G.D., *Brains and Reasoning* (London: Macmillan, 1974), 132, and elsewhere in that book for closely related remarks.
22 Cf. op. cit. note 5, section 5.13.

23 Beckerman, W., 'The Problem of Judging Evil', *The Times*, London (17 December 1986), 18; see also, by the same author, *The Times*, London (7 February 1987), 8.
24 Marciano, W., and Pagels, H. 'Quantum Chromodynamics', *Physics Reports* (Section *C* of Physics Letters) **36** (1978), 137-176.
25 Lewontin, R.C., *The Genetic Basis of Evolutionary Change* (New York: Columbia University Press, 1974).
26 Laplace, P.S., in *A Philosophical Essay on Probabilities*, 6th edn, trans. by F.W. Truscott and F.L. Emory (New York, 1961), 4, argued:
Given an intelligence which could for one instance comprehend all the given forces by which nature is animated and the respective situations of the beings who compose it - an intelligence sufficiently vast to submit these data to analysis - it would embrace in the same formula the movements of the greatest bodies of the universe and those of the lightest atoms, for it nothing would be uncertain, and the future, as the past, would be present to its eyes.
27 Cf. Vernon, M.D., *A Further Study of Visual Perception* (Cambridge University Press, 1952); A. Michotte, *Bulletin de l'Academie Royale Belgique* (class des Lettres) **34** (1948), 268ff.
28 Levin, B., 'Programmed to Choose', *The Times*, London (20 January 1987), 14.
29 Popper, K.R., 'Natural Selection and the Emergence of Mind', the First Darwin Lecture given at Darwin College, Cambridge, 8 November 1977, and K.R. Popper, in Popper, K.R. and Eccles, J.E., *The Self and its Brain* (Berlin: Springer International, 1977).
30 Wassermann, G.D., 'Reply to Popper's Attack on Epiphenomenalism', *Mind* **88** (1979), 572-575, (or see chapter 6 above).
31 Hayek, F.A., *The Sensory Order* (London: Routledge and Kegan Paul, 1952).
32 The Archbishop of York, Dr John Habgood, 'Man's Freedom with Responsibility', *The Times*, London (12 January 1987), 12.
33 Kolb, E.W., Seckel, D., and Turner, M.S., 'The Shadow World of Superstring Theories', *Nature*, London **314** (1985) 415; see also J. Ellis, 'The Superstring: Theory of Everything, or of Nothing?', *Nature*, London **323** (1986) 595-598.
34 Cf. Wassermann, G.D. 'On a Physical (Materialistic) Theory of Psi-phenomena based on Shadow Matter', *Inquiry*. **31** (1988) 217-233; see also Wassermann, G.D., *Shadow Matter and Psychic Phenomena* (Oxford: Mandrake of Oxford, 1993).

16 'Morality and determinism': a reply to Flew

Antony Flew's criticisms[1] of my paper 'Morality and Determinism'[2] seem to me flawed because of (1) Flew's disregard of the existence of extensive statistical (probabilistic) determinism in the natural sciences and (2) Flew's out-of-context quotations of parts of my sentences with resulting misrepresentations of my meanings. I wish to substantiate this.

Flew alleges[3] that Beckerman and I share wholeheartedly the premise of total and *necessitating* determinism. Yet, *pace* Flew I wrote[4] that 'In many branches of science the scientific systemic modelling involves an idealized statistical determination. For instance, in quantum mechanics the 'equations of state' of systems provide scientific models which only yield a statistical [probabilistic] determination.' Thus, quantum mechanical behaviour of many systems is implicitly probabilistic and not necessitating. Flew's claim that 'the world of the natural sciences knows nothing of any but necessitating causes'[5] is therefore invalid. Again, Flew claims[6] that 'true laws of nature necessarily necessitate'. Yet such laws, e.g. Boyle's law or Ohm's law, are empirically obtained by fitting curves to data by statistical procedures.[7] The fitted curves do *not* necessitate where data points have to be located relative to the curves. Thus both the social *and* the natural sciences depend explicitly and/or implicitly on statistics, in particular on probabilities, and do not causally or otherwise necessitate anything in Flew's sense.

Flew also cites only *parts* of some of my sentences, and thus out of context. This occurs when Flew claims that[8]

> Thus he [Wassermann] allows that 'many psychopaths may be genetically strongly predisposed' and that 'many psychopathic personalities could have a genetic basis for their personality defect' without ever asking whether this genetic predisposition or this genetic basis is to be hypothesized and

understood as guaranteeing that any subsequent psychopathic behaviour will be outwith the psychopath's control.

Instead, I had written[9] 'Although, for all we know, many psychopaths may be genetically strongly predisposed, *only further research can clarify this issue.* Currently it remains unknown whether most or even much psychopathic behaviour has a genetic basis.' Again, I wrote[10] 'I conclude at this stage that although many psychopathic personalities *could* have a genetic basis for their personality defect, *no reliable information exists...concerning any possible genetics of such disorders, and such disorders are also difficult to diagnose*'. Thus, Flew omitted twice my strong doubts about the extent of a genetic basis of much psychopathic behaviour. With such doubts it would have been futile for me to speculate what the possible genetic basis of some psychopathic behaviour could achieve (as Flew thinks I should have done, see above). Flew also misrepresents my views elsewhere.[11] I argued[12] that '...human beings could, perfectly well, be sentient self-organizing automata whose brains are developmentally wired in so as to have the *capacity* to think that they have "free will"... The outcome of this capacity may be deceptive and an illusion. Simply because people like Levin think that we have "free will" we cannot be sure that this alleged "free will" is other than make-believe. In fact, one might ask "by what criteria would Levin's *mentalism*[13] guarantee that 'free will' is something apart, some purely mental entity, which is not a product of matter?" ' Misleadingly, Flew only cites the preceding sentence, omitting the previous three sentences (in defence of materialism) and asserts that the only sentence he quotes is rhetorical, which, only *out of context,* is the case.

After misrepresenting me, Flew asserts[14] that 'it is possible and, surely, better to argue that both the human subject and the responsible agent are one and the same creature of nothing else but flesh and blood and bone? Agents, when and inasmuch as they are agents, must be able to do other than in fact they do.' Yet, *pace* Flew, this applies equally to a hydrogen atom which from its ground state could go into any one of a variety of permissible excited states, i.e. the atom has a 'choice' of alternative states. Which state is selected depends then on *environment*al excitation. Similarly, a sentient automaton, whether called an 'agent' or something else, could, under appropriate environmental influences, assume, with the same machinery, alternative states. This does not mean that there exists indeterminacy in the system, but there could exist quantum mechanical statistical determinacy. Flew writes as if 'agents', when they assume alternative states, do so autonomously. Yet, environmental influences could possibly have made the difference. Finally, Flew attributes fictitious arguments to me[15] which I have neither put forward nor would advocate, because of the *statistical* determinism which, according to our quantum mechanical models, rules over all bodily ions, atoms and molecules (etc.).

Notes

1 Flew, A., ' "Morality and Determinism": Two Comments', *Philosophy* **64** (1989), 98-103.
2 Wassermann, G.D., 'Morality and Determinism', *Philosophy* **63** (1988), 211-230. (Reproduced in Chapter 15 above with different pagination)
3 Op. cit. note 1, 98.
4 Op. cit. note 2, 224.
5 Op. cit. note 1, 99.
6 Op. cit. note 1, 99.
7 Guest, P.G., *Numerical Methods of Curve Fitting* (Cambridge University Press, 1961).
8 Op. cit. note 1, 100.
9 Op. cit. note 2, 217.
10 Op. cit. note 2, 219.
11 Op. cit. note 1, 101.
12 Op. cit. note 2, 227.
13 See op. cit. note 2, section 2.
14 Op. cit. note 1, 101.
15 Op. cit. note 1, 101-102.

17 Wittgenstein On Jews: some counter-examples

Wittgenstein's empirical generalizations about the personality structure of Jews, in *Culture and Value*,[1] as cited and reconstructed by Yuval Lurie,[2] seem to me to be formally and factually of the same misguided genre as many untrue empirical generalizations about Jews in Richard Wagner's essay *Judaism in Music*, in the (forged) *Protocols of the Elders of Zion*, in Hitler's *Mein Kampf*, in the arch-anti-Semitic work by the late Nazi Nobel Laureate Philipp von Lenard, *Deutsche Physik*[3] (which claimed that one must distinguish Jewish from Aryan Physics), and in myriads of other anti-Semitic tracts produced before, during, or after the Nazi period. Although Lurie (op. cit. note 2) gave an excellent *exposition* of Wittgenstein's defamatory generalizations about Jews, he did not *refute* them philosophically and factually. I shall, therefore, try to provide such refutations.

Wittgenstein who, by his own admission, was three-quarters Jewish (op. cit. note 2, p. 342), considered himself as a Jew (op. cit. note 1, p. 21). As a Jewish self-denigrator he is not unique, and his views on Jews appear to be modifications of the anti-Semitic generalizations of the originally Jewish, but baptized, Otto Weininger (op. cit. note 2, pp. 328 ff), who was a dedicated admirer of the vile anti-Semite Richard Wagner. (Self-hatred of members of minorities is a common phenomenon; see the article on Jewish Self-hatred in the *Encyclopaedia Judaica*, Year Book, 1977-78, pp. 351 ff..). Unfortunately, Wittgenstein accepted uncritically, and propounded dogmatically, as if true, some anti-Semitic views that were already widely current in Germany and Austria in the 1920s. He relied, usually implicitly, on *empirical generalizations* of the type '*All* Jews have personality trait values *X, Y, Z...*' i.e., empirical generalizations of the type '*All S are P*'. On purely *philosophical* grounds I have argued elsewhere[4] that such non-statistical empirical generalizations are faultily formulated and lead to *Hempel's Paradox*. They should be replaced by statistically formulated empirical generalizations, which avoid Hempel's Paradox. Indeed, even older publications on the psychology of personality structure[5] showed already that within a specific population, such as the non-Jewish population of either France or Britain, or the Jewish population of the world, there exists considerable interpersonal variation of each personality trait value, leading to a statistical distribution of each value for that population. For instance, a certain percentage of non-Jewish Britishers are introverts, another percentage are extroverts, and similar remarks apply to Jews and other populations. Again, some non-Jewish Frenchmen and some Jews may be cunning, while other members of the same populations may be straight, and there may be intermediate shades. This variability of each personality trait, within known populations, completely rules out stereotyped empirical generalizations that rely on single-value, non-variable, personality traits for a population. Yet, Wittgenstein relied precisely on such fictitious one-value personality traits, for instance when he invoked the obscure concepts of *the* 'intellect' or *the* 'character'. Thus, Wittgenstein uses empirical generalizations that are invalid in form *and* content (see below). The intra-population variation of personality traits demands statistically formulated generalizations, which do not make assertions about *all* members of a population. Wittgenstein apparently failed to recognize this, and this is the major reason for the inappropriateness of most, or all, of his statements about Jews.

Typically, Wittgenstein argued that 'even the greatest of Jewish thinkers is no more than talented'...and...'that the Jewish mind does not have the power to produce even the tiniest flower or blade of grass; its way is rather to make a drawing of the flower or blade of grass which has grown in the soil of another mind and to put it into a comprehensive picture'.[6] Put differently, Wittgenstein seems to suggest that although some Jews (including himself) are talented, *all* Jews lack originality and genius, which are non-Jewish traits only (op. cit. note 2, p. 331) a view that was also shared by most Nazis. The preceding quotation from Wittgenstein illustrates his use of inappropriate forms of (implicit) empirical generalizations in relation to personality traits, by disregarding variability of these traits. Undoubtedly, many Jews, like many non-Jews, are not particularly original and have no genius, whereas relatively few Jews, and relatively few non-Jews, are highly original and display genius. By admitting only a single value per (e.g. Jewish) population for each personality trait variable, Wittgenstein created what Lurie, on the basis of arguments differing from mine, called an abstract, platonic, 'metaphysical stereotype of a Jew'. Lurie regards such personality stereotypes as 'socially reprehensible and philosophically picayune' (op. cit. note2, p. 336), and my preceding arguments provide additional reasons why Wittgenstein's stereotype of a Jew is inappropriate, and many reinforcing arguments for this will follow.

(In the sequel numbers given in parentheses will refer only to page numbers in op. cit. note 1.).

The fictitious Wittgenstein stereotype of a Jew is supposedly: cunning (22), secretive (22), and, although he may be clever and learned, has to base his affairs on nothing, since he lacks a foundation (19). He can only be reproductive (19), i.e. unoriginal and imitative (19), and, hence, cannot display originality and genius (except when a 'holy man', whatever that may mean (18)). The Jew, lacking originality and genius, has at best only talent (19). He is a 'tumour' in the 'tissue' of the non-Jewish populations among whom he resides (20), lusting for money as a means to attaining power (21), but lacking an appreciation of property possession for its own sake (21). The Jew of the stereotype also has an 'intellectual' (13) but 'characterless' spirit (35) (since according to Wittgenstein (35) 'the measure of genius is character' and as he assumes that, except for holy men, Jews have no genius, they have no character). This 'characterless' Jewish spirit makes use of the spirit of host nation members (19) to contribute to *their* culture. Jews, however, cannot create a culture of their own (since by Wittgenstein's *hypothesis* they lack the required originality, and can only be reproductive (19); see also op. cit. note 2, p. 329).

In the, just cited, numerous Wittgenstein empirical generalizations about Jews, which are all hypotheses, the personality trait variable values, selected arbitrarily by Wittgenstein as applying to Jews, form a conglomerate of negative extremes. By selecting almost only *negative* personality trait values and allocating *all* of them to Jews, Wittgenstein succeeds in constructing a caricature, namely a rare, or even fictitious, Jew as a stereotype. For instance, to imply, as Wittgenstein does, that Jews lack 'character' must appear as absurd to any psychologist of personality structure, since everybody has a 'character', and character can vary widely across the same population, and simply cannot have 'zero' value for every member of the same (Jewish) population. Wittgenstein also believes that the 'intellect' of Jews allows them to contribute to civilization, although 'intellect' unlike 'character' cannot, in his view, produce originality and genius, only talent.

Wittgenstein's *metaphysical* stereotype of a Jew is difficult to reconcile with the *facts* concerning a multitude of *creative* Jewish people, who produced works of great *originality* and *genius*, as judged by numerous criteria of excellence, and who contributed much cultural novelty in the sciences, the arts, philosophy, engineering, mathematics, music (etc.). Far from being a 'tumour' within their host societies, these Jews, and many other Jews who by their enterprise built up businesses that provided work for many non-Jews and produced many important goods, proved to be of benefit to the host societies. The reason why there are anti-Semites in many countries, who make faulty empirical generalizations about Jews, is a matter for the *psychology of prejudice*, which has been widely studied by social psychologists (cf. G.W. Allport, *The Nature of Prejudice* (Reading, Mass: Addison Wesley, 1954)). According to Allport, anti-Semitism has more to do with the personality structure of anti-Semites than that of the Jews they hate, and he characterized some typical personality trait values that appear to turn up dominantly in anti-Semites, anti-Coloureds (etc.).

I now propose to cite some counter-examples to Wittgenstein's allegations that Jews are not original and devoid of genius. My examples all relate, as far as I know, to (ethnic) full-Jews, since for partial Jews (e.g. Half-Jews) of genius Wittgenstein could have argued that their genius was possibly due to their 'non-Jewish' inheritance. This forced me to omit a number of celebrated people who are half-Jewish, and even to omit Wittgenstein himself from my list of Jewish philosophers. Let me first of all cite some of the Jewish *Nobel Laureates*.[7] These include *in Physics*: Felix Bloch; Max Born; Albert Einstein; Richard P. Feynman; James Franck; Murray Gell-Mann; Sheldon L. Glashow; Donald A. Glaser; Brian Josephson; Lev Davidovich Landau; Gabriel Lippmann; Robert Hofstadter; Benjamin R. Mottelson; Arno A. Penzias; Isidor Isaac Rabi; Julian Schwinger; Emilio Segré; Otto Stern; and Steven Weinberg. (Other eminent and original Jewish physicists include Sir Rudolf Peierls; E. Teller; Lise Meitner; Herbert Fröhlich and Emil Wolf (former president of the American Optical Society).)

Jewish Nobel Laureates *in Medicine and Physiology* include: Julius Axelrod; David Baltimore; Robert Bárány; Ernst Boris Chain; Stanley Cohen; Konrad Bloch; Baruch S. Blumberg; Gerald Edelman; Paul Ehrlich; Joseph Erlanger; Sir Bernard Katz; Sir Hans Krebs; Karl Landsteiner; Rita Levi-Montalcini; Joseph L. Goldstein; Otto Meyerhof; Daniel Nathans; Tadeus Reichstein; Howard Temin; Selman A. Waksman; Marshall W. Nirenberg; George Wald; Salvador Luria; Rosalyn Yallow; Fritz Lipmann; Joshua Lederberg; Arthur Kornberg and Otto Loewi. (Other prominent and original Jewish bioscientists include the *evolutionary biologists* Ernst Mayr; Richard C. Lewontin and Eviatar Nevo, and the *developmental biologists* Paul Weiss and Lewis Wolpert and elsewhere in Biology Sir Walter Bodmer.)

Jewish Nobel Laureates *in Chemistry* include: Paul Berg; Melvin Calvin; Fritz Haber; Gerhard Herzberg; Francois Jacob; Aaron Klug; André Lwoff; Max Perutz; Richard Willstaetter and William Howard Stein.

Jewish Nobel Memorial Laureates *in Economic Sciences* include: Milton Friedman; Franco Modigliani; Paul Samuelson; Leonid Kantorovich; Simon Kuznets; Kenneth J. Arrow.

Jewish Nobel Laureates *in Literature* include: S.Y. Agnon; Henri Bergson; Saul Bellow; Nelly Sachs; Isaak Bashevis Singer and Boris Pasternak. (There exist many other famous and original Jewish writers such as Franz Werfel; Arnold Zweig; Stefan Zweig; Arthur Schnitzler; Lion Feuchtwanger; Jakob Wassermann;

Arthur Koestler; André Maurois; Norman Mailer; Philip Roth; Bernard Malamud; Frederic Raphael; Elaine Feinstein and Louis Golding among others.)

So we have here 69 Jewish Nobel Laureates, showing, *pace* Wittgenstein, that the Jews have contributed their quota of originality and genius in the fields of Physics, Medicine, Physiology. Molecular Biology, Chemistry, Literature and Economics. By genius I mean here the *Concise Oxford Dictionary definition* of 'extraordinary imaginative, creative or inventive capacity.' Contrary to the massive evidence cited above (and below) that there exist Jews of genius (who are not holy men), Wittgenstein had the temerity to imply that 'genius is a quality to which men of Jewish spirit never attain' (op. cit. note 2, p. 331 and (18)), an obviously absurd and totally wrong empirical generalization.

When it comes to distinguished *original* Jewish *philosophers*, there are, among others, Yehoshua Bar-Hillel; Max Black; Spinoza; Sir Isaiah Berlin; Hermann Cohen; Herbert Feigl; Philipp Frank; Edmund Husserl; Stephan Körner; Nelson Goodman; Arthur Pap; Sir Karl Popper; Moritz Schlick; Alfred Tarski; Friedrich Waismann, Morton White and, whether one agrees with him or not, Karl Marx (cf. Siegmund Kaznelson, *Juden im Deutschen Kulturbereich*, (Berlin, Jüdischer Verlag, 1959) and *Encyclopaedia Judaica* for some of the philosophers listed). (The late Sir Alfred Ayer was half-Jewish).

Eminent and *original* Jewish *Mathematicians*, many of *genius*,[8] include: Binyamin Amira; Abram Besicovich; Sir Hermann Bondi (Applied Mathematics); Abraham Adolf Fraenkel; Richard Courant; Izrail Moisevich Gelfand; Aleksander Gelfond; Jacques Solomon Hadamard; Kurt Hensel; Sidney Goldstein (Applied Mathematics); Walter Hayman; Felix Hausdorff; Adolf Hurwitz; Karl Gustav Jacobi; Samuel Karlin (Statistics, particularly as applied to evolutionary science); Felix Klein; Leopold Kronecker; Edmund Landau; Solomon Lefschetz; Tullio Levi-Civita; Jacob Levitski; Rudolf Otto; Zigismund Lipschitz; Hermann Minkowski; Richard von Mises (Statistician); Sir Claus Moser (Statistician); Louis Mordell; Johann von Neumann (who ranks among the greatest mathematicians of the twentieth century); Emmy Noether; Werner Rogosinski; Klaus Friedrich Roth (Number Theory); Isaiah Schur; Laurent Schwartz; Beniamino Segre; Hugo Dyonizy Steinhaus; James Joseph Sylvester; Otto Toeplitz; Fritz Ursell (Applied Mathematics); Vito Volterra; André Weil and Norbert Wiener (founder of Cybernetics). Although there exists no Nobel Prize in Mathematics, there exists a corresponding distinction for mathematicians, namely the Fields Medal (founded in the 1930s and awarded to pure mathematicians below the age of 40). Prominent *Jewish Fields medalists* include: Paul Cohen; Charles Feffermann; Klaus Friedrich Roth and G.A. Margulis (USSR). It is extraordinary that Wittgenstein should have implied, *ex cathedra*, that Jewish Nobel Prize winners and Fields medalists could not possibly be people of genius because they are of 'Jewish spirit' (18) and ethnically Jewish.

There were also numerous *original* and prominent Jewish *Painters* and *Sculptors*. Jewish painters of distinction include:[9] Jankel Adler; Yosl Bergner (Israel); Marc Chagall; Felix Glück; Jozef Israels; Ernst Josefson; Isaac Lewithan; Max Liebermann; Amedio Modigliani; Felix Nussbaum (killed at Auschwitz); Camille Pissaro; Sir William Rothenstein; Victor Weiss (a famed cartoonist known as Vicky). Well known *original* Jewish *Sculptors* include: Benno Elkan; Sir Jacob Epstein and Chaim Gross. Many people regard, at least, Chagall as a genius, and this rating is likely to extend to some others of those named.

Eminent Jewish *Composers*[10] of considerable *originality* and genius include (*pace* Wittgenstein's mudslinging at Mendelssohn (2), (35), (38), (39) and Mahler (67)): Mendelssohn (e.g. his incidental music to Shakespeare's Midsummer Night's Dream, his Octet, his violin concerto in E minor, his symphonies and much else); Gustav Mahler (e.g. his nine symphonies) and Arnold Schoenberg. Other prominent Jewish composers of originality (some also conductors) include: Leonard Bernstein; Leo Blech; Aaron Copland; Paul Dukas; Hans Eisler; George Gershwin; Karl Goldmark; Jaques Halévy; Emmerich Kálmán (composer of operettas); Erich Korngold; René Leibowitz; Giacomo Meyerbeer; Darius Milhaud; Jacques Offenbach; Kurt Weill (of *Dreigroschenoper fame);* Jaromi Weinberger (composer of *Schwanda the Bagpiper*) and Irving Berlin (composer of musicals).

Jewish conductors and instrumentalists, who gave *original* interpretations, include among *Conductors*: Gustav Mahler; Leonard Bernstein; Bruno Walter; Otto Klemperer; George Szell; Sir George Solti; Pierre Monteux; André Previn; Lorin Maazel; Daniel Barenboim; James Levine; and include among *Instrumentalists*: David Oistrakh; Jascha Heifetz; Sir Yehudi Menuhin; Itzhak Perlman; Pinchas Zuckerman; Henryk Szeryng; Emil Gilels; Murray Perahia; Arthur Schnabel; Arthur Rubinstein; Vladimir Horowitz and Vladimir Ashkenazy.

Typical among Wittgenstein's many unjustified *racist* empirical generalizations is the assertion that 'The Semitic races have an unpoetic mentality, which heads straight for what is concrete' ((6) and op. cit. note 2, p. 323). Here, again, Wittgenstein ought to have realized that the personality trait variable of 'being poetic' varies widely over a population. Some Jews are very poetic, some mildly poetic, others non-poetic, and the same applies to non-Jews. For instance, Heinrich Heine was a famous German-Jewish poet, and Hayyim Nahman Bialik was a great Hebrew poet, thus refuting Wittgenstein's claims, once more, by counter-examples, and refuting the almost identical claim of the virulently anti-Semetic composer Richard Wagner that Jews cannot write poetry.[11] Thus, Wittgenstein absorbed like a sponge, and re-emitted, anti-Semitic lies, past and present. According to Wagner, the Jew 'speaks all the time like a foreigner' (op. cit. note 11, p. 57). Yet, surely, this does not apply to Bialik who wrote in Hebrew, the Jewish national language! Shabtai Teveth has published the following extract from a letter by David Ben-Gurion to Sir Arthur Wauchope (former British High Commissioner to Palestine):[12]

Bialik was our greatest poet since the Destruction, and apart from the Prophets and Psalmists he was the all-time master of the Hebrew language... He was the spiritual leader of his people, the master craftsman of the revival of our language. No Jew of our time has had such a powerful impact on the life of his people... In our eyes he was the bearer of the nation's legacy, of its ideals and aspirations. Fate has struck us in taking Bialik, the greatest Jew of our time, all too soon.

Other Jewish poets of eminence (some of whom are also famed as novelists) include: Ossip Mandelshtam (a renowned Soviet-Jewish poet, who, among more important work, wrote a satirical poem about Stalin, which led to his murder in a Soviet camp);[13] Boris Pasternak; Isaak Rosenberg (a British Jew); Elaine Feinstein (British-Jewish); Franz Werfel; Jehuda Amichai and Hayyim Lenski. The preceding list of celebrated or distinguished Jewish poets shows that Wagner's and Wittgenstein's empirical generalizations regarding Jews and poetry, are simply false. Wagner knew of Heine (op. cit. note 11, Chapter 9) and should have known better, and Wittgenstein's life partly overlapped that of Bialik and Mandelshtam, and he must have heard of Heine (who was much more famous than Wittgenstein). Hence, there is no excuse for his deplorable and unjustified generalizations about Jews and poetry. Presumably, he ignored Jewish poets, notably Heine, Bialik and Mandelshtam, in order to convince himself that, true to his hypothesis, Jewish poets of eminence did not exist and could not be men of genius, just as he dismissed Mahler's music as worthless (67) and asked whether Mahler should not have committed suicide rather than have written such bad symphonies (67). Among famous and original Jews I must also mention Noam Chomsky, one of the most prominent theorists of linguistic structure.

When it came to wrong generalizations about Jews, Wittgenstein was, as shown, partly paralleled by Wagner. Wagner strongly influenced Hitler, the worst mass-murderer of Jews in history. According to William L. Shirer (*The Rise and Fall of the Third Reich* (London: Book Club Association, 1960), p. 133) Hitler used to say 'Whoever wants to understand National Socialist Germany must know Wagner.' Wagner, as we saw, was also deeply admired by Weininger (cf. op. note 11, p. 59), who, in turn, strongly influenced Wittgenstein (19), as we saw, so both Hitler's and Wittgenstein's views are partly, directly or indirectly, rooted in Wagner. At least Hitler's views were also strongly influenced by the vile anti-Semite Houston S. Chamberlain, the son-in-law of Wagner (cf. Shirer op. cit. above). Although, in *Culture and Value*, Wittgenstein never mentioned the

Holocaust, he stated that 'It isn't sensible to be furious even at Hitler; how much less so at God' (46) (ca. 1945), an odd point of view, to say the least.

Wagner alleged that Jews are especially deficient in the realm of vocal music.[14] Yet, Mendelssohn wrote the oratorio *Elijah* (etc.) and Mahler composed *Das Lied von der Erde, Lieder eines fahrended Gesellen, 12 Lieder aus des Knaben Wunderhorn* and many other fine songs. Schoenberg composed the *Gurrelieder* and the opera *Moses and Aaron*; Offenbach composed the *Tales of Hoffmann* and many other vocal works; Bernstein wrote *Westside Story*; Gershwin composed *Porgy and Bess*; Weinberger wrote the opera *Schwanda the Bagpiper* and Kurt Weill composed the *Dreigroschenoper*. There is also Halévy's opera *La Juive* (recently recorded for the first time); Hans Eisler's work *Die Massnahme*; Emmerich Kálmán's operetta *Countess Maritza*, and there exists a good deal of recorded vocal music by Copland. In the field of musicals Irving Berlin is eminent. Thus, *as composers of vocal music Jews are far from deficient*. In case of any lingering doubts readers should listen to Mahler's partly vocalized symphonies Nos. 2, 3, 4 and 8 which, in the opinion of many people, are works as great as many or any Wagner wrote, although symphonies and operas are hard to compare. *Pace* Wagner, there are also well known Jewish singers, who do not indulge, as Wagner put it (see op. cit. note 11, p. 57), in synagogue-based gurgling, yodelling and prattling.

These singers include: Evelyn Lear (soprano), Beverly Sills (soprano); Friedrich Schorr (bass-baritone, a famed singer of Wotan in Wagner's 'Ring'; something Wagner did not live to know, since he died in 1883, while Schorr was born on 2 September 1888; I had this confirmed by Miss M. Worswick who has a taped talk by the musicologist William Mann on Schorr; Mann states that Schorr was the son of a Jewish cantor); Richard Tucker (tenor); Robert Merrill (baritone); George London (née Burnstein, bass-baritone); Joseph Schmidt (tenor); Alexander Kipnis (bass); Selma Kurz (coloratura soprano); Regina Resnik (contralto, a well-known opera singer) and Sir George Henschel. (Most of these are listed in op. cit. note 10 and in the *Encyclopaedia Judaica*, Year Book, 1974, pp. 367-376). Also Richard Tauber (tenor, famed Lieder and operetta singer) is widely believed to be Jewish, although he is not mentioned in the *Encyclopaedia Judaica*, which, however, does not mention Schorr either, although he was certainly Jewish, and is mentioned as Jewish in the *Oxford Companion to Music*.

We can add to the preceding list of cultural achievements, often of great originality, often of much genius, by many Jews, also important *original* achievements of Jews as technological inventors. Thus, Emil Berliner (inventor of the disc record); Robert von Lieben (inventor in the field of communication engineering); Siegfried Marcus (prominent in the development of combustion engines); Leon Arons (inventor of the mercury vapour lamp) and Bruno Strauss (inventor of stainless steel) are but a few among those cited in *Encyclopaedia Judaica* (under 'Inventors and Engineers'). There is also the notable British-Jewish inventor Paul Eisler (of printed circuits fame) (cf. Association of Jewish Refugees (London) *Information* **45**, No. 2 (1990), 7).

I conclude that, *pace* Wittgenstein's assertions about the complete lack of originality (i.e. only 'reproductiveness' (19)) and genius of Jews, a large number of Jews have contributed, often with extreme originality and genius to many branches of culture and civilization, and that the Jews have produced fine and original composers, singers, conductors, instrumentalists, scientists, inventors, poets, novelists, painters, sculptors and philosophers. Thousands of years before Wittgenstein lived, the Jews had already in their own country a flourishing culture of their own, created by them, with their own Hebrew language, and their own form of worship. Yet, Wittgenstein implies (see above and (19)) that Jews cannot create a culture of their own (cf. op. cit. note 2, p. 329). Writing as he did, Wittgenstein was obviously out of his depth, forgetting the most obvious facts, and disregarding or dismissing others that did not suit him (e.g. dismissing Mahler's music as worthless (67)). Just as the late virulent anti-Semite von Lenard tried in a futile way to distinguish Jewish Physics from Aryan Physics, so Wittgenstein thought, equally absurdly, that one must 'distinguish between works of culture that stem from a Jewish spirit and works of culture that stem from a non-Jewish spirit; ((19) and op. cit. note 2, p. 329). This generalization is again false. One can certainly say that the Hebrew language, the Jewish form(s) of worship and religious celebration, the Yiddish theatre and Yiddish language and literature as they existed in parts of Eastern Europe before the Holocaust, are *typical* of specific Jewish cultures created by Jews. But, above this Jews (like non-Jews who have their own language and other cultural aspects specific for them) share in many cultural aspects that are not specifically Jewish but international. One might ask 'what is 'Jewish' about Einstein's theory of the photo-electric effect for which he received a Nobel Prize?' The answer is: there is, obviously, nothing specifically Jewish about this theory, any more than the theory of quantum mechanics, initiated by the celebrated non-Jews Heisenberg, Schrödinger and Dirac, can be said to be 'non-Jewish'. Similar remarks apply to a host of other Jewish creative work.

Again, *pace* Wittgenstein, who claimed that Jews have no appreciation of property possession ((21) and op. cit. note 2, p. 324), this is, once more, a wrong empirical generalization. Many Jews, like many non-Jews, have a great love of property. They collect art treasures, stamps, recorded music, and buy or build houses for themselves, and sometimes build up excellent libraries (etc.). Equally wrong is Wittgenstein's generalization that Jews desire money as a means for attaining power (21) and not as a kind of property. The alleged Jewish urge for power was widely proclaimed by anti-Semites in Germany, Austria, Imperial Russia and elsewhere, and the allegation came from the forged *Protocols of the Elders of Zion*. The 'Protocols' alleged that there exists a Jewish world conspiracy to take over all sources of power in the world, and for Jews, thereby, to become masters of the world. (This looks more like Hitler's aspiration, and was very nearly realized by the Germans in world war II. In fact, a favourite (pre-war) Nazi song went 'Today we own Germany, tomorrow the world'). Jews often love property not because it confers power, but because of pride of ownership. Jews like Lord Mond (founder of ICI), Lord Hirst (founder of GEC, UK), Sir Eric Weiss (founder of Foseco-Minsep, UK, now with many branches abroad), Sir Jules Thorn (founder of Thorn Electrical Industries, U.K.), and many other Jews in the UK and in many other lands, built up industrial enterprises that gave employment to many non-Jews and that greatly helped to enrich the world's civilization. There is no evidence that this industrialization was done because of a lust for 'power' rather than for enjoyment of *ownership of such property* and enterprise (*pace* Wittgenstein).

I have now demonstrated, by citing numerous counter-examples and by noting the formally wrong structure of Wittgenstein's empirical generalizations about Jews, that these generalizations are *totally* wrong. Lurie (op. cit. note 2) while exposing Wittgenstein's views on Jews, did not, in my opinion, sufficiently invalidate these views. I accept that Wittgenstein was a philosopher of genius. Yet, it was people who argued like him about Jews, that helped to lay the foundations of the Holocaust, in which some of my friends and family perished.

I am indebted to Dr Alex Rubner (London), Miss Marjorie Worswick and Dr James Ford (Newcastle University) for supplying me with some important items that helped in my compilation of counter-examples. I am also grateful to the staff of the Wiener Library (4 Devonshire Street, London W1), notably Mr A. Wells, for making available numerous works and some names that assisted in the compilation.[15]

Notes

1 Ludwig Wittgenstein, *Culture and Value*, G.H. von Wright (ed.) translated by Peter Winch, (Oxford: Basil Blackwell, 1980).
2 Lurie, Y., 'Jews as a Metaphysical Species', *Philosophy* **64** (1989), 323-347.
3 von Lenard, P.E.A., *Deutsche Physik* 4 vols. (München: J.F. Lehmann, 1936-37).
4 Wassermann, G.D., 'On the Laws of Nature', *Synthese* **51** (1982), 381-396.
5 Cattell, R.B., *Personality* (New York: McGraw-Hill, 1950).

6 Op. cit. note 2, 323; and op. cit. note 1, 18 and 19.

7 The names of these Nobel Laureates and brief descriptions of their works can be found in T. Wasson (ed.), *Nobel Prize Winners* (New York: W.H. Wilson Company, 1987); see also under their names in the *Encyclopaedia Judaica*.

8 This listing of Jewish Mathematicians is partly based on the *Encyclopaedia Judaica* (under the heading 'Mathematicians').

9 Révész, H., *Jewish Painters*, unpublished but available at the Wiener Library (London); see also Karl Schwarz, *Jewish Artists of the 19th and 20th Centuries* (New York: Philosophical Library, 1949).

10 Révész, H., *Jewish Composers and Musicians*, unpublished but available at the Wiener Library (London); see also Gdal Salenski, *Famous Musicians of Jewish Origin* (New York: Bloch Publishing Co., 1949), and *The Oxford Companion to Music* (under Jewish Musicians).

11 Martin van Amerogen, *Wagner, A Case History* (translated by Stewart Spencer and Dominic Cakebread) (London: J.M. Dent & Sons, 1983), 57.

12 Shabtai Teveth, *Ben-Gurion* (London: Robert Hale, 1987), 456-457. Dr Rubner pointed out the quotation.

13 Cf. the *Independent* (newspaper, London), 9 December 1989, 31 for an article on Mandelshtam.

14 Op. cit. note 11, 57.

15 My compilation of Jews of originality is not intended to be exhaustive, but selective and illustrative.

PS. Mention must also be made of the distinguished Jewish archaeologist Brian Shefton (cf. Association of Jewish Refugees (London) Information 51, No 1 (1996), 2). (Added in the proof of this book.) Further there is a famed Jewish Nobel Laureate Joseph Brodsky (literature and poetry). His obituary is in *The Times* (London) 30.1. 1996, p. 17.

18 Moderate relativism and psychological sets

18.1 Introduction

This paper aims to propound and defend a variety of moderate relativism (MR) which avoids extreme relativist claims and which partly relies on the notion of 'psychological set' (PS). The paper is also intended as a contribution to a prominent ongoing debate between certain relativists and anti-relativists, notably in the social sciences.[1] First I shall briefly consider the *notion* of PS. Much has been written about PSs by various psychologists[2], and F.H. Allport summarized some of their main properties[3]. Attitudes are typical PSs, and PSs can act as *determining tendencies*. Indeed, cognitive sequences and related actions are not programmed and enacted randomly[4], but may be determined by dominant PSs which could persist for many years. For instance, Richard Wagner is reputed to have persisted for more than 25 years in constructing his music dramas of 'Der Ring des Nibelungen'. PSs are often, perhaps always, goal-directed and, conversely, every human teleological activity, whether long-term planning, or near-automatic walking along a street towards a 'pre-set' destination, depends on PSs. (Biochemically PSs could be due to long-term or short-term sensitizations of brain machinery that represents cognitive constructs and planned related actions[5].) In their most extreme forms PSs seem to be involved in simple enactments of posthypnotic suggestions, and hypnotic states could be interpreted as being due to abnormally induced and retained PSs. Above all, many PSs *can be learned*[6] so that, for instance, many of our attitudes can be learned.

My brand of MR claims that each of the variety of social groups to which any one individual belongs overtly, or by tacit agreement with their views or actions, tends to establish in that individual PSs specifically related to the views and actions of that social group. Accordingly my MR asserts that many of our opinions and actions are partly or completely conditioned by our group-related PSs. These

PSs are not necessarily immutable (although this may apply to some PSs of very 'rigid' personalities), but newly acquired PSs may sometimes replace in dominance existing PSs as determinants of beliefs and actions. This happens, typically, in religious conversions, or when we acquire a system of PSs relating to a new scientific theory. For a moderate relativist of my persuasion, however, it is essential to realize that *some of the PSs of many, or most, human beings are similar to the PSs prevalent among members of particular groups of people.* The groups concerned may be social groups, such as particular populations of a nation (say the Scots or Welsh), or the adherents of a particular religious belief, or an ethnic group, or experts in a branch of learning (e.g. physicists who accept a particular hypothetico-deductive theory as 'useful' and applicable), or political parties, and so forth. Any individual might be endowed with PSs relating to various groups to which that person belongs (e.g. a Roman Catholic could be an expert in some branch of theoretical physics). Some people's PSs may cause them to *believe* that some views or statements are universally true, when, in fact, one is only dealing with group-confined *'accepted truths'* (ATs). ATs are *alleged* 'truths' that are accepted by members of a particular group (e.g. religious doctrines). Acceptance of a 'truth' by a group does not imply universality of that truth, since other groups may not accept it. As Newton-Smith put it[7]

> The relativist is impressed by the variation in belief and in reasoning from age to age from social group to social group. This variation, the more extreme relativist tells us, arises from the fact that the different ages or groups live in different worlds. Different things are true for them...

This extreme relativist claim, however, is exaggerated and not embraced by MR of the kind expounded here and elsewhere[8]. Two people *A* and *B* could be members of social groups *E, F, G, H* and *I* (for *A*) and *E**, *F, G, H* and *I* (for *B*) respectively and could thus *share* many of the PSs and ATs relating to groups *F, G, H* and *I*, but could differ with respect to ATs and PSs relating to groups *E* and *E**. Thus a Roman Catholic mathematician and an orthodox Jewish mathematician may differ with respect to the ATs of their respective *religious groups* but may embrace the same set of ATs of pure mathematics in their specialized fields of mathematics. This casts doubts on Gellner's claim[9] that

> A spectre haunts human thought: relativism. If truth has many faces, then not one of them deserves trust or respect...

This alleged 'cognitive anarchy' (op. cit. p. 108) seems to me to be exaggerated. Different ATs accepted by adherents of different religions do not cast doubt on, or lower respect for, the ATs of a particular branch of mathematics.

The present MR is not just concerned with comparing primitive with advanced societies (in the fashion of Winch[10]). This MR is equally concerned with opposed belief systems and ATs within so-called 'advanced' societies and the PSs that they generate. Typically irreconcilable groupings within Western (and other) societies are mechanistic materialists versus interaction-dualists, or creationists *vis-à-vis* evolutionists[11,12,13] or opposed political parties. To sum up, the MR here defended claims that our beliefs, whether metaphysical or scientific (etc.), and our ATs, depend very much on our group-related PSs.

18.2 Inter-Group transmissibility of truths

Newton-Smith (op. cited note 7, p. 108) argued that if a sentence *S* is true in language A of tribe T, then if *S'* is a synonymous translation of *S* into another language used, say, by social group G, *then S'* must be true for *G*. This, however, only applies to *universally true* sentences (see also Gellner's op. cited note 9, which emphasizes universalism vis-à-vis relativism) and not to ATs which may be social group-specific. Does this liquidate MR? Not at all, since my type of MR asserts that many of the most prominent truth claims of Western culture are *not* universal truth statements, but ATs accepted by particular social groups. For instance, the axioms of various pure mathematical theories are not provably true, otherwise they would not be axioms. Even where one can show that axioms are consistent this does not prove their universal truth. Thus, mathematical axioms may only be considered as ATs *relative to an appropriate group of expert pure mathematicians.* If, however, the axioms of a mathematical theory are *assumed* to be true, in the sense of ATs, then the *provable* truth of certain derived theorems follows. Yet, since the axioms have only AT status, the theorems derived from the axioms cannot be universally true, but can only be ATs for the group of experts that assume the truth of the axioms (since one cannot deduce universally true statements from statements that are only ATs without demonstrating the universal truths of these ATs). It remains possible that tautologies are the only universally true statements. Certainly the starting hypotheses of scientific hypothetico-deductive theories (e.g. Newton's first, second and third 'laws' of mechanics) can neither be proved to be universally true nor do they have AT status, since scientists accept that these hypotheses are very fallible as regards their range of applicability.

Various philosophers have established several *theories* about the nature of truth (e.g. the 'correspondence theory' of truth). Yet, these theories themselves depend on certain *basic assumptions,* which are not provably universally true. Thus, while any adherent of a particular truth theory can state what he or she *means* by the (notion of) truth, such persons cannot demonstrate the universal validity of their truth theory. Like any other theory, a particular truth theory can, as a theory, only claim AT status relative to the group of those who accept it.

18.3 Contingent truth claims

Many contingent truth statements, such as assertions which claim the truth of some fact (e.g. 'it is true that this specimen of grass looks green') *are accepted* by 'normal' (e.g. non-colour-blind) members of all cultures. Acceptance of the contingent truth (CT) of many everyday statements depends, probably, on the fact that human nervous systems are normally, via genetic control of development, basically similarly constructed (subject to considerable interindividual variations; cf. op. cit. note 4a), and, accordingly, represent many things or events in the world around us in essentially similar ways. CTs, however, are ATs, although the group of people who accept a CT may, for reasons just stated, often comprise most of mankind. Indeed it would be wrong to suppose that *widely accepted CTs* are universal truths. To see this let us consider an example from R. Gregory[14] (op. cit. p. 161) who noted (in a different context) that

> Studies of people living in dense forests have been made. Such people are interesting, in that they do not experience distant objects, because there are only small clearances in their forests. When they are taken out of their forest and shown distant objects, they see these not as distant but small. People living in Western cultures experience a similar distortion when looking down from a height. From a high window objects look small, though steeplejacks and men who work on the scaffolding and girder structure of skyscrapers are reported to see objects below them without distortion.

(Such examples suggest that mechanisms akin to, or of the same kind as, those involved in PSs may be operative even in adjusting by *learning* to some of the simplest ways of perceiving the world.) Gregory's example shows that although it may be a CT (and an AT) for inhabitants of dense forests that 'distant objects (shown to them) are small', this need not be a CT or an AT for groups of people who normally do not inhabit dense forests. It therefore may be unwarranted to claim, as some philosophers do (e.g. an Oxford don in a private communication), that there exist 'contingent and surely uncontrovertible truths'.[15] As Gregory's example shows, some CTs are only ATs for particular human groups, although such groups may often comprise most of mankind. For instance, some kinds of visual illusions cannot be perceived by some people, or only with great difficulties. These, and other, similar, examples, familiar to psychologists (cf. note 15), reinforce the view that our nervous systems may function similarly for most, but not necessarily all, people and that certain alleged CTs are possibly ATs for many, possibly most, but not necessarily all, people provided that they are endowed with 'sufficiently normal' brains and other physiological systems. Indeed, people's PSs can cause them to adopt CTs or ATs *relative* to small or large groups, as the case may be. I need not repeat here the views of all other moderate or extreme relativists and anti-relativists, since I wish to restrict myself mainly to discussing the views of some particularly prominent anti-relativists that merit special attention in the present context[16]. Whereas the axioms of mathematics (and their deductions) may be interpreted as ATs, rather than as universal truths, the starting hypotheses of hypothetico-deductive theories of science (e.g. Newton's three 'laws' of classical mechanics or Maxwell's equations of electromagnetic theory) are not even regarded as 'truths' in any sense by many scientists. In addition to their starting hypotheses hypothetico-deductive theories, when applied to particular 'systems', invoke often *idealized* system-specific *ad hoc* hypotheses (e.g. that strings are 'light' or massless and 'perfectly inelastic' and that some bodies are 'smooth' are among the many system-specific *ad hoc* hypotheses often invoked in elementary classical mechanics). Any scientist knows that 'idealizations' are simplifications which are not intended to be accepted as 'truths' (either as CTs or ATs), but they are often helpful in formulating 'approximately valid' models of systems.

18.4 Group beliefs

Tribal beliefs, when translated into English, may, in some cases, appear bizarre, or even unintelligible, to English readers, but so do many aspects of the general theory of relativity or of quantum mechanics, when presented to a non-expert man in the street. Experts in quantum mechanics form as distinct a social group as the upholders of a tribal belief or of any other religious belief. To an educated English speaker the belief system of the Fataleka of the Solomon Islands may seem peculiar, and what is an AT for the latter may not be an AT for the former. According to Sperber[17,18] the Fataleka.

> maintain not only that the earth is flat, but also that it is the fifth of nine parallel strata among which various entities are distributed: a person's reflection is in stratum three, flutes in stratum four, crocodiles in stratum seven, stratum eight is empty...

But then, within Western societies, we have many evolutionists who may find many claims of creationists as hard to accept as the beliefs of the Fataleka. For instance, Marks[19] writes

> The creationists claim that we may come to know the Creator by deducing Him from his works on Earth. Well, we know that there are 750,000 species of insects now alive. Anyone who has ever had a conversation with an entomologist knows what a deadly dull lot those fascinated by insects can be. To envision the Creator expending his divine energy on three quarters of a million meticulously grafted bugs necessitates an image of Him as a dismal insufferable cosmic bore!

Yet, within the same Western societies there are creationists who may share the views of Professor Paul Rosenbloom of Columbia University (op. cit. note 13) that

> Rashi's commentary on the sixth day of creation suggests the possibility that God created the entire world during the first six days and put everything in its proper place. He then set it all in motion when he made Adam. The hypothesis that God created the universe complete with fossils, rocks, etc., as they should have been 5743 years ago, is consistent with all scientific laws. The choice between this hypothesis and one like the Big Bang Theory is a matter of faith, not of scientific fact.

The preceding examples illustrate striking differences of ATs accepted by different groups within Western societies.

Lukes[20], who is primarily concerned with 'alien beliefs of tribal and traditional societies, and pre-modern scientific theories' has cited the so-called 'Principle of Charity' according to which 'one should be maximally charitable in assigning truth conditions across indicative sentences of the language which are held true by those being interpreted.'[21] This principle can certainly be applied to ATs of primitive as well as advanced societies. On the other hand I do not agree with the prominent thinker Davidson's[22] claim (quoted by Lukes op. cit. note 20) that we can

> take it as given that *most* beliefs are correct. The reason for this is that a belief is identified by its location in a pattern of beliefs; it is this pattern that determines the subject-matter of the belief, what the belief is about. Before some object in, or aspect of, the world can become subject-matter of a belief (true or false) there must be endless true beliefs about that subject-matter.

In the sciences, for instance, although many *facts* may be provisionally regarded as ATs, at least until improved observational techniques lead to revised ‘factual claims’, the *idealizations and approximations* that are adopted, when applying hypothetico-deductive theories of the sciences to the formulations of specific models of particular systems, can hardly be claimed to yield ‘true’ beliefs or ATs (see Section 18.3). Our scientific theories and their hypotheses may have immense pragmatic value which, however, does not entitle us to regard them as ‘correct’ or ‘true’ beliefs, because they are always liable to revision or to being superseded by improved hypotheses. Although good scientific theories are immensely *valuable* as regards insights and applications, this does not confer on these theories the status of ‘truths’.

18.5 Do some societies think illogically?

I agree with Lukes (op. cit. note 20) that Lévy-Brühl’s view, that ‘primitives live and act in a pre-logical and mystical world’, is grossly wrong, and that there exists ‘no evidence for different *kinds* of reasoning processes’ among primitive people (except for schizophrenics (etc.) among primitives). Despite this, advocates of ‘cognitive relativism’ assert, as Sperber (op. cit. note 17) put it, ‘that people of other cultures live “in other worlds”, so that what is rational in their world may well appear irrational in ours.’ Certainly some culturally established PSs of primitive people might *simulate* illogicality or irrationality of pronouncements when these are heard or read by members of more advanced societies. Yet, many members of advanced societies are prone to occasional (or even frequent) irrationalities. A *capacity for rational* thinking of primitive people is shown by their ability to deal in terms of tribal legal reasoning with topics such as house building, fishing, navigation, trade and exchange (as noted by Lukes, op. cit. note 20, p. 270). Yet, use of this capacity in such contexts does not rule out irrationality elsewhere. For instance, allegedly ‘advanced’ mediaeval European societies, who also had competent legal systems for dealing with the range of affairs just listed, in addition had many members who believed in witchcraft (a group-bound system of PSs concerning beliefs). I conclude that although similar (though interindividually varying) *capacities* for rationality may be species-specific for the species *Homo Sapiens,* acquired PSs can cause different groups of our species to acquire beliefs and to exhibit actions which appear irrational to other groups. Adolf Hitler’s system of PSs were hardly superior to many of the PSs of members of extremely primitive societies.

18.6 Consensus based on accepted standards

By definition, ATs (and in particular CTs) depend on *consensus* among members of the appropriate social groups that accept such truths (see Sections 18.1 and 18.3). That most truth claims concern ATs rather than 'universal truths' is the *essence* of the present MR. This MR suggests also that social, scientific and many other problems have solutions that are only accepted on the basis of *agreed standards* of appropriate social groups. These standards (e.g. moral standards, or standards of 'rigour' in mathematics etc.) can change as time goes on. Standard-based consensus may be based on plausibility arguments and justificatory arguments, or on simple aesthetic appeal of intellectual creations, such as works of art, music and poetry.

Essential causes for accepting particular belief systems (and their ATs and standards) rather than others are often educational, and depend on the educational standards of the groups we are either born into or choose to belong to. Moreover, some of our acquired PSs may prejudice us against views of other cultures or ideologies or the skin colour of members of other societies. Political propaganda may generate strong PSs which may be hard to eradicate, manifesting themselves as dogmatism. According to certain psychologists some people may be genetically endowed so as to be more inflexible than others, thereby retaining their PSs, and, hence, their belief systems, more strongly in the face of overwhelming evidence to the contrary[23]. The apparent existence of strong PSs, and dogmatism caused by them, in some individuals makes the advocacy of 'critical rationalism' by Popper, Bartley and others very utopian (cf. discussion in op. cit. note 8). Although Bartley[24] and Popper advocate that we should display clearly our positions and expose them to maximum criticism, people with 'strong views' (due to persistent powerful PSs) are often unlikely to be swayed by justified criticisms and will continually evade criticisms, where possible by ingenious counter-arguments. That PSs can be powerful is indicated by the effects of posthypnotic suggestions. There are, however, no good reasons for believing that all our PSs are acquired culturally (by learning) and some PSs may be innate (e.g. many of our perceptual tendencies, revealed by Gestalt psychologists, *appear* to be innate, but see my remarks on inhabitants of dense forests in Section 18.3 above).

Because of the likely existence of (perhaps many) innate PSs I reject Sperber's claim (op. cit. note 17, p. 155) that

> From a relativist point of view, then, all conceptualized information is cultural. What we think of as the sky, birds, eyes, tears, hunger, death, fall in other cultures under concepts which differ from our own, and are therefore perceived differently... Propositions that can be entertained, expressed, asserted are, according to relativists, language - and culture-specific...

This statement, however, simply does not reflect the point of view of the present MR. Because of innate colour-perceiving retinal- and brain machinery people of all cultures (who are not colour blind) may claim that the sky appears to be blue to them when this can be objectively verified by physical measurements (etc.).

18.7 Facts and beliefs

It is well known that scientific observations often depend on theories of the observing instruments used, and, hence, on hypotheses, so that even 'facts' are hypotheses-linked. As noted above, facts, like hypotheses can at most be subjects of contingent truth statements but not of provably universally true statements (and I regard even mathematical axioms as only acceptably true relative to the group of people who accept them for purposes of further arguments). Whereas physicists can at least *classify* facts and starting hypotheses of hypothetico-deductive theories in different ways, although accepting the contingent nature of truths claimed for facts and/or theories; anthropologists are sometimes in doubt whether a statement of a member of a tribe is a fact or a belief. Only the skill of practitioners of the art can help to resolve these issues.

18.8 Liquidation of psychological arguments against relativism

Among the various arguments against relativism, most of which are easy to dismiss, Sperber (op. cit. note 17, p. 159) raises one objection to relativism which merits discussion in order to dismiss it. He states

> Imagine an organism capable of developing the cognitive abilities of either the cat or the dog, depending. say, on whether it was raised among cats or among dogs. For this, it would need to possess innate abilities sufficient to match those of either species, plus some extra device capable of determining in which of the two cognizable worlds it had landed. It takes richer innate capacities to learn to be a cat or a dog than to be either. In the case of humans (as seen by relativists) the surplus of innate capacities required in order to determine the right cognizable world would be incommensurably greater since there are not two, but an infinity of profoundly different accessible worlds, each of great complexity.

The fallacy of this argument is not hard to detect. For instance, a human child has (and needs) only *one innate capacity* for language acquisition, since that capacity enables it to learn not only the languages of any society it happens to be born into, but, in the case of a polyglot, permits to acquire two or more languages using the same innate capacity. More generally, man has innate capacities to learn most varied things. For instance, a gifted physicist can learn not just one, but several, or perhaps all, of a number of immensely complex known hypothetico-deductive theories (of the kinds enumerated near the end of Section 18.3), thus using the *same innate capacity* to master many different theories. Hence, the potential (innate) capacity for learning cognitive constructs of certain classes outstrips in most cases the actual use made of that capacity. For instance, Jordan[25] notes that the celebrated pianist Walter Gieseking, playing invariably from memory, gave concerts filling whole evenings with piano works which he had never played before, but only studied while travelling on trains. Obviously, the same innate capacity could have been used for learning alternative or additional piano works. Or, again, the British journalist Bernard Levin[26] states that

> Given the right cross-references I can recite whole pages of books I have not opened for 30 years, and I have sometimes astonished and alarmed my friends by quoting remarks of theirs that go back decades.

These and other accounts show that man has a vast innate surplus capacity for learning the most varied things.

The preceding examples, rather than Sperber's fictitious cat-dog creature seem to speak against Sperber's assertion (op. cit. note 17, p. 159) that

> A relativist model of development would have to represent a much more complex process and, *ceteris paribus,* to rely more heavily on innatist hypotheses than a universalist one. The usual argument against universalism, that it applies unnecessary assumptions about innate mechanisms, should actually weigh and quite heavily against relativism.

Yet, the only innatist assumption required by MR is that the general capacity to learn a great variety of things (some stated above) is innately acquired. Assuming the capacity for learning languages to be innate, as it undoubtedly is, is not the same as assuming that we are born with an innate universal grammar, as some linguists have repeatedly assumed. I do not think that any developmental psychologist would deny that the capacity for learning languages is innate, although which languages we learn depends on environmental contingencies. *Our innate capacity to learn does not exclude that we learn the cultural items of particular groups, many of whose PSs, and much of whose acquired knowledge we may learn.* Sperber argues as if a relativist had to assume that our *actual knowledge* is innate, whereas MR *only has to assume that our capacity to acquire knowledge, and not knowledge per se, is innate.*

Universalists believe that there exists only one intellectual world. Against this Gellner (op. cit. note 9, p. 200) aptly concluded that

> The solution of the problem of relativism does not hinge on the establishment of human universals. If it has a solution, it lies elsewhere. Relativism is about the existence of One World: and the conceptual unification of the world is, precisely, the work of one particular style of thought, which is not universal amongst men, but is culturally specific. But this in turn does not actually subvert the Universality Thesis: for although the conceptual unification of the world does have specific socio-historic roots, it is evidently accessible to all men, and is in fact now being diffused generally.

Gellner's claim that 'the conceptual unification of the world...is evidently accessible to all men' involves, however, some wishful thinking. Most members of primitive societies (or groups) and even many members of advanced societies simply have very little, or any, contact with the more sophisticated products of the 'conceptual unification of the world'. Moreover, the innate capacities of (normal) members of the species *Homo Sapiens* to acquire complex cognitive repertoires vary considerably. Some people find it hard to learn simple school mathematics, let alone the most sophisticated branches of mathematics that exist, and similar remarks apply *mutatis mutandis* to complex musical compositions, advanced theoretical physics and much else. Thus, the membership of *expert social groups* (and access to their PSs) will always remain confined, in many cases, within rather narrow limits, just as those who master more than one or two languages form small subgroups of various populations. This suggests that the basis of relativism is likely to persist and seems rooted partly in limitations of innate capacities of many individuals, which excludes them from actual or potential membership of certain groups (e.g. of experts), and it also may be rooted in lack of opportunities of access to advanced products of conceptual unification.

Defenders of MR may be right in one important respect. A metaphysical belief that the nature of the reality of the material structure of the universe can be (at least partly) represented by human cognitive constructs seems to be well founded. Yet, these constructs are fallible scientific theories whose universal truth we cannot prove (even, if by chance, some of these theories were universally true). Instead we must be content with the (often) temporary acceptance of scientific theories that supposedly model the real material world, and that are produced by experts. These models, instead of being provably universally true, or even contingently true, are only valid *relative* to the group of experts who proclaim, understand and accept them. What seems almost 'miraculous' is the *progress of science* which appears to be *converging* towards an ever deeper understanding. To committed mechanistic materialists, like myself, it looks as if human brains have evolved so as to provide materially encoded guidance systems that *collectively*, get ever closer to optimal explanations of the structure and function of material systems ranging from quarks to brains themselves. Thus, the *capacity* for contributing to this progress seems to be wired into many human brains developmentally (under genetic control) although the opportunity for contributing may be restricted by particular group membership and other contingencies.

18.9 Ranking of group beliefs

It is one thing to agree with advocates of MR that, despite the existence of certain universal goals, there exist many different groups with group-specific belief systems, PSs, customs rituals etc. Yet, racists, and others often assert the *superiority* of the belief system of one group over that of another group. Some racists regard their own group as being superior to that of other 'inferior' groups (or races) and the holocaust of Jews was instigated by the Nazis precisely on the grounds that they, as 'Aryans', were a group superior to the Jews, and that the Nazis had to rid the world of 'inferior' races. This type of barbaric and perverse relativism must, of course, not be confused with the MR discussed here, since this MR is *not normative* and does not assert absolute ranking of groups, or prescribe elimination of any groups. Thus, relativists need not be Nazis or ultra-Marxists (who believe in the 'liquidation' of their opponents, or in the maltreatment of members of other groups). If the Inquisition is anything to go by, then, with its burning and torture of infidels, it provided a classic example of gross intolerance of one group towards adherents of other groups. Yet, my advocated MR does not preach group intolerance. My brand of MR is a doctrine based on *observations* of group-related PSs. Some of these PSs may further human intellectual progress, while others may lead to the destruction of much human life. Yet, my kind of MR, while trying to analyze how some aspects of human rationality or irrationality arise (notably via adherence to PSs), does *not proscribe* deliberate attempts to change people's PSs by critical debate à la Popper (rather than sending them to labour camps), even if a truly 'Open Society' (in Popper's sense) may be an utopia.

18.10 Summary

A moderate relativism (MR) is expounded and defended. It is argued that some of our 'psychological sets' (PSs) are partly or entirely determined by the views of social groups to which we belong either overtly or by tacit agreement. PSs are apt to generate strong prejudices in favour of certain belief systems and to overrule considerations of alternative views. Popper's 'critical rationality' is suggested to be utopian since it underrates the strength of PSs. The views of Sperber, Lukes and other partial, or strong, anti-relativists are examined and criticized. It is also stressed that even the *notion* of truth is relative to certain group-accepted 'theories of truth'. Some 'universalist' claims (e.g. those of Gellner) seem to be based on over-optimism and the claims of the present MR seem to be closer to practically encountered situations than these universalist claims.

Gerhard D. Wassermann was until recently Reader in the Theory and Philosophy of Biology in the University of Newcastle upon Tyne, with which he has still close ties. He is the author of *Molecular Control of Cell Differentiation and Morphogenesis* (New York: Marcel Dekker, 1972), *Brains and Reasoning* (London: Macmillan, 1974), *Neurobiological Theory of Psychological Phenomena* (London: Macmillan, 1978, and a Japanese edition) and author of numerous papers in scientific and philosophical journals. Within philosophy he has centred on philosophy of science and philosophy of mind.

Notes

1 Cf. M. Hollis and S. Lukes (eds.), *Rationality and Relativism,* Oxford: Basil Blackwell, 1982.
2 E.g. D.O. Hebb, *The Organization of Behavior,* New York: John Wiley, 1949; F.A. Allport, *Theories of Perception and the Concept of Structure,* New York, John Wiley, 1955.
3 Op. cit. note 2.
4 Wassermann, G.D., *Neurobiological Theory of Psychological Phenomena,* London: Macmillan, 1978, p. 185.
4a Wassermann, G.D., 'The Wiring-in of Neural Nets revisited', *Bulletin of Mathematical Biology,* **48**, 1986, pp. 661-680.
5 Op. cit. note 4, p. 36.
6 Allport, op. cit. note 2.
7 Newton-Smith, W., 'Relativism and the Possibility of Interpretation'. In op. cit. note 1, p. 106.
8 Wassermann, G.D., 'Social Aspects of Rationality', *Methodology and Science* **17**, 1984, pp. 50-69.
9 Gellner, E., 'Relativism and Universals'. In op. cit. note 1, p. 181.

10 Winch, P., 'Understanding a Primitive Society', *American Philosophical Quarterly* **1**, 1964, pp. 307-324.

11 Cf. the book review by R.K. Bambach, 'Responses to Creationism', *Science* **220**, 1983, pp. 851-853, which lists and discusses extensively recent anti-creationist views. See also G.D. Wassermann 'On the Nature of the Theory of Evolution', *Philosophy of Science* **48**, 1981, pp. 416-437 for one recent exposition of basic evolutionary tenets.

12 Cf. Wassermann, G.D., 'Testability of the Role of Natural Selection within Theories of Population Genetics and Evolution', *British Journal for the Philosophy of Science* **29**, 1978, pp. 223-242; G.D. Wassermann, 'Kindly explained', *Nature Lond.*, **296**, 1982, p. 284; G.D. Wassermann, 'Reply to Popper's Attack on Epiphenomenalism', *Mind*, **88**, 1979, pp. 572-575; G.D. Wassermann, 'Materialism and Mentality', *Review of Metaphysics* **35**, 1982a, pp. 715-729; G.D. Wassermann, 'Quantum Mechanics and Consciousness', *Nature and System*, **5**, 1983, pp. 3-16.

13 Rosenbloom, P., 'What we cannot know', *B'or Hatorah* (Israel) No. 3, Winter-Spring, 1983, pp. 5-11. I am indebted to Professor E. Nevo, Director of the Institute of Evolution of Haifa University for drawing my attention to Rosenbloom's paper.

14 Gregory, R., *Eye and Brain*, London: Weidenfeld and Nicolson, 1966.

15 Hochberg, J., In *Woodworth and Schlosberg's Experimental Psychology* (3rd edn.), J.W. Kling and L.A. Riggs (eds.), London: Methuen, 1971.

16 Cf. op. cit. note 1.

17 Sperber, D., 'Apparently Irrational Beliefs', In op. cit. note 1, pp. 149-180, p. 154.

18 Sperber bases this account on R. Guidieri, *La Route des Morts*, Paris: Seuil, 1980, p. 47.

19 Marks, J., 'All in the book', *Nature Lond.* **295**, 1981, p. 276.

20 Lukes, S., 'Relativism in its Place', In op. cit. note 1, pp. 261-310.

21 See D. Davidson, 'Psychology as Philophy' in D. Davidson, *Essays on Actions and Events*, Oxford: Clarendon Press, 1980, pp. 238-9; D. Davidson, 'On the very idea of a conceptual scheme', *Proceedings of the American Philosophical Association* **47**, 1973-4, p. 19.

22 Davidson, D., 'Thought and talk'. In S. Guttenplan (ed.), *Mind and Language*, Oxford: Clarendon Press, 1975, p. 20.

23 Allport, G.W., *The Nature of Prejudice*, Reading, Mass.,: Addison-Wesley, 1954.

24 Bartley III, W.W., 'The Philosophy of Karl Popper, Part III: Rationality, criticism and logic', *Philosophia* **11**, 1982, pp. 121-221.

25 Jordan, P., *Verdrängung und Komplementarität* (2nd edn.), Hamburg: Bergedorf Stromverlag, 1951, p. 89.
26 Levin, B., *Sunday Times*, 9 November 1983, p. 33.

Note

This paper was first published in *Methodology and Science* (1987) **20**, 223-237.

19 Quantum mechanics and consciousness

I

In this paper I shall discuss one common, but widely accepted, fallacy, and another less familiar fallacy concerning the relation between human consciousness and quantum mechanics. The first fallacy was propounded in a book review by Marie Jahoda in *Nature*. She writes:[1]

> The revolution in the basic assumptions of physics brought about by Einstein and by the development of quantum theory has repercussions far beyond that discipline. One of the consequences of the new physics - that the mind of the observer is a necessary ingredient of the structure of theory - has not only returned terms like 'mind' or 'consciousness' to respectable scientific status; more far reaching in the views of growing numbers of scientists, philosophers and historians of science it also undermines the mechanistic world view that originated with Bacon, Descartes and Newton.

I shall argue that these conclusions simply do not follow if one embraces a defensible kind of materialism, such as epiphenomenalism[2] or the closely related 'materialistic dualism' which I have recently espoused[3] and which some people regard as being close to a dual aspect theory.

Jahoda asserts that 'people did not and do not regard themselves as machines notwithstanding the existence of robots.' This allegation is probably applicable to the man in the street, and to those who attach themselves to 'curious sects, pseudo religions, creationism, blind imitations of Eastern practices' and so forth. Yet, it does not apply to many hard-faced mechanistic materialists, whose point of view I shall try to defend. There is no *a priori* reason as to why certain classes of machines should not be conscious, without thereby losing their status as machines. Unless we define the concept of 'machine' to comprise only certain unconscious systems, there is no reason why there should not be conscious machines. Perhaps what worries people is the kinds of systems human beings are, particularly their immense physiological and behavioural adaptability to environmental givens, which *prima facie* seems so very different from the types of systems that we find in factories or in our homes and which we term 'machines'. So let me first of all say something about automata and their mechanisms.

II

To start with I shall quote a very relevant passage from a recent paper dealing with the nature of the theory of evolution[4]. I wrote:

> In general a mechanism relates to one or more systems. A 'system' could be a mechanical system such as a top, or a biological system such as a member of the species *Drosophila melanogaster*, or a population of a species, to mention but a few among a host of possible examples. The theory of a mechanism - if it exists and is properly formulated - is in its general aspects a hypothetico-deductive theory which is usually applicable to mechanisms of a wide class of systems, and which, subject to additional system-specific *ad hoc* hypotheses, explains how mechanisms of particular members of that class of systems operate. Known theories of mechanisms are manifold. They include theories of mechanisms of mechanical, electrical and biological systems, and so forth. For instance, in the life sciences there are mechanisms of DNA replication, protein synthesis, photosynthesis, muscle contraction, nerve impulse propagation etc.

In the paper, from which I just cited, I proceeded to list the hallmarks of a typical hypothetico-deductive theory of a mechanism, and I shall not repeat here the details listed, since I have also cited and discussed them at great length elsewhere.[5] Suffice it to stress here that quantum mechanics is a typical theory of mechanisms, dealing with such 'systems' as atoms, ions, molecules, fundamental particles (such as quarks in quantum chromodynamics and leptons in the theory of weak interactions). Anybody who has studied any of the many existing texts on quantum mechanics will have noticed that the notion of consciousness of the observer enters neither directly nor indirectly into the axiomatics of quantum mechanics.[6] I shall return to this issue later in this paper.

Automata (or machines) are particular classes of systems. Some people would regard any atom or molecule as an automaton, and there are no good *a priori* reasons for excluding these simple systems from the general class of automata, except, perhaps, classificatory arguments. Once we accept atoms and molecules among the class of automata, then we might be tempted to classify any structure built up from atoms or molecules - including mountains and houses and oceans - as 'compound automata'. This, undoubtedly goes beyond what most classifiers would accept within the category of automata.

When scientists and philosophers debate whether organisms and their subcomponents (such as brains) are automata, confusion must be avoided by distinguishing different kinds of automata. First of all there are automata such as automobiles, television sets, aeroplanes, and so forth, with *fixed numbers* of components. These machines can change the states of their components (a sparking plug of a car engine, for instance, may be in a sparking or non-sparking state) but not the types or numbers of components. Electronic computers, for instance, are relatively inflexible, and their apparent versatility lies in the variety of man-made programs which can be fed into them (although self-programming computers have been considered). Turing[7] established a general theory for certain classes of fixed component machines,[8] and electronic computers of certain kinds are special cases of Turing machines. The program on which a computer system operates provides detailed instructions for stepwise execution of an algorithm or other kind of program.

I must now turn to another class of automata, namely *self-organizing systems*. They have the property that their component types and numbers are not fixed from the start, and that new component types can be generated by the automaton during its functioning or development. For instance, the development of higher organisms involves the formation of myriads of new cells and new cell types[9] and the formation of memories by brains requires the building up of new engrams. Wound healing demands regeneration of missing cells and/or cell types. In a self-organizing automaton some components, once established, could remain permanently fixed, while others undergo repeated modifications. The important point is that Turing machines are not self-organizing automata, but some self-organizing (and other) automata can in some respects function like Turing machines or be controlled by them. (Although the *output* of a computer may control industrial production of new parts of another machine, in doing so the computer does not change its own structure.)

From the preceding arguments it is obvious that people cannot be regarded as machines in the sense in which aircraft, cars, television systems, digital computers, and so forth are considered as machines. Yet, there are no *a priori* reasons against interpreting living organisms as automata, namely self-organizing automata. An ever increasing number of studies by life scientists consistently confirm that the functioning, structure and behaviour of organisms is amenable to explanations in terms of theories of mechanisms. Hence from the biochemical and biophysical point of view there are no valid arguments available to suggest that people could not regard themselves as self-organizing automata, whose modes of self-organization can be fully comprehended in terms of physical theories, such as quantum mechanics, electromagnetic theory, and so forth. While these arguments suggest that people may be self-organizing automata, they still do not answer the question whether the theories by which we describe such automata are such that the mode of self-organization depends on consciousness, or whether people are describable as self-organizing automata that, because of the way in which they have organized themselves - without the intervention of consciousness in the self-organizing processes - happen to be constructed so as to be capable of consciousness. I shall argue, as I have done elsewhere[10] that the latter remains an important *a priori* possibility (of a metaphysical kind), and that for this reason Jahoda's principal argument breaks down. It must also be added, before I proceed, that even if many people for sentimental reasons do not regard themselves as automata (even as conscious automata), this does not imply that they could not be automata (endowed with consciousness) just the same. After all, many people would not frequently think of themselves as having descended from *Homo erectus.* The status that people desire to assume has no *bearing* on the nature of their origin or on their material basis, and, hence, on their actual status among the world of material systems.

III

I have recently propounded a modern version of epiphenomenalism (or as some people might prefer to call my theory, a modern version of a 'dual aspect' theory[11]). This theory is purely metaphysical and presents an important *a priori* possibility, and to start with I shall briefly recapitulate some of its essential features, leaving the reader to study the detailed arguments, related to the materialistic dualism to be expounded, in my recent paper referred to.

Everyday matter is built up from elementary particles, namely two kinds of quarks and electrons. (An elementary particle is supposed to be structureless, as revealed, for example, by scattering experiments with other constituents of matter.) The interactions between the elementary particles of everyday matter are mediated, as appropriate, by various postulated interaction-mediating constituents, the evidence for some of these force-mediating entities being in various cases only very circumstantial, although very convincing. These postulated entities are gluons (mediating strong forces), intermediate vector bosons (mediating weak and electromagnetic forces) and gravitons (supposedly mediating gravitational forces). Interaction-mediating material constituents are theoretically predicted by quantum theories of appropriate interaction fields.[12] Some, or all, of the types of elementary particles as well as the types of interaction-mediating material constituents could be possible 'mental-state-associated material constituents' (MSAMCs). By hypothesis MSAMCs have physical state parameters associated with them (such as mass, spin, charge, colour) as well as having associated conscious states, when appropriately activated. Conversely, materialistic dualism hypothesizes that all conscious states are states of MSAMCs and that conscious states cannot occur independently of MSAMCs. Whether MSAMCs are identical with the hitherto known elementary particles and/or force-mediating material constituents, or whether MSAMCs are subcomponents of these elementary particles and/or force-mediating material constituents can never be known with certainty for several reasons. First it cannot be assumed that any conscious states are necessarily associated with a single particle or a single force-mediating constituent, since it remains possible that any one conscious experience is associated with an ensemble composed of more than one elementary particle and/or force-mediating material constituent. Hence, the term MSAMC could refer to such an ensemble rather than to a single material entity. Second, the behavioural criteria of human consciousness do not apply to MSAMCs (since they cannot talk or cry, etc.). For these two reasons alone materialistic dualism must remain no more than a metaphysical *a priori* possibility. Materialistic dualism asserts that conscious states are irreducible to physical states (i.e., that conscious states are inexpressible in terms of physical state parameters), and that there exists at present no parametric language for describing conscious states. Even if someone invented a parametric language for characterizing conscious states, the conscious state descriptors would be distinct from physical state parameters, according to materialistic dualism.

According to materialistic dualism all physical interactions between elementary particles, and all interactions due to force-mediating material constituents are in principle completely describable in terms of the language of physical state parameters and reference to conscious states is never required when describing these interactions. It is also assumed by materialistic dualism that a conscious state can never have any direct or indirect influence on any material constituents. A conscious state can occur if and only if a specific type of physical state of an appropriate system of MSAMCs occurs. It is disclaimed that a physical state of a system of MSAMCs ever 'causes' any conscious state (or vice versa), since causal languages and causal processes of the familiar kinds do not apply to the relationships between physical and conscious states. Although each conscious state has an associated matter state, the relationship between the two kinds of states cannot at present - and quite likely never - be symbolized in any but trivial formulations of the type 'A conscious state of type X is associated with activities A of physical systems (e.g. specified parts of brains) of type B.' Materialistic dualism also postulates that a conscious state of a material system M cannot causally interact with the conscious state of any other material system M'. It is only when M and M' interact causally in a manner that is in principle describable by their physical state variables that conscious states of systems M and M' may occur which may appear to be associated.

It follows from the preceding assumptions of materialistic dualism that conscious states cannot occur spontaneously without concomitant changes of physical states of material systems. Hence, the occurrences of conscious states are not autonomous but are always linked to the occurrences of matter states or of transitions of matter states. By contrast changes of physical states of matter need not necessarily in all cases be accompanied by changes of conscious states, although this possibility cannot be ruled out. Moreover, since all causal interactions between material systems are describable in terms of physical variables, any interactions between physical systems will - unless physics should demand otherwise - conserve the energy of the interacting (or reacting) systems. In particular, according to materialistic dualism, all 'actions' (of people) only require physical state variables to describe the changes involved that are revealed by behaviour and detectable brain states, although more holistic ways of describing actions are usually adopted. Certain specific changes of brain states that are associated with human actions can, however, be associated - in particular - with conscious states. Those conscious states that occur concomitantly with the action - causing physical states of matter could then convey the feeling that they have 'caused' the actions, although this feeling would be spurious and misleading.

There are, as is well known, many philosophers of mind, who, unlike myself, advocate a 'materialistic monism' which invokes some form or another of a contingent mind-state/ brain-state (or mind-event/ brain-event) identity theory.[13] It seems to me that any contingent identity theory would need, over and above the postulates of materialistic dualism, assumptions concerning the nature of the contingencies envisaged. It would have to state in each specific case which particular concatenation of brain state parameter values corresponds to a specific conscious experience, so that one could regard this correspondence as a contingent identity. This requirement looks to me to be adding a great deal more metaphysics to materialistic dualism than is necessary. One would only be justified in proclaiming a contingent identity as a general hypothesis if it could be shown that for different brains of different human beings similar brain states would lead to similar conscious experiences. Yet this would at once create two-fold trouble. First, different people's brains may, because of the presence of gene variants (alleles) in different individuals, be endowed with different variants of a considerable number of proteins. If consciousness should depend on states of proteins, say - and nobody could rule this out with certainty - then different people simply could never be expected to have identical brain states. This conclusion applies also for many other reasons (e.g. differences in the number and size of neurons of a particular type in the brains of different individuals). Second, even if people claim to have similar experiences, their assurances are only linguistic tokens of alleged experiences of consciousness, and thus, do not guarantee any (logical) identity of experiences. Hence, in the case of a contingent mind-event/brain-event identity statement the precise terms of reference on either side of the contingent identity can never be precisely known and stated. It is for these and other reasons of difficulties of specifications (notably of unambiguous knowledge of relevant brain states) that I prefer a materialistic dualism to any form of hitherto proposed (metaphysical) 'contingent identity theory.'

IV

This concludes the essential recapitulation of my metaphysical theory of materialistic dualism, which appears to be coherent and consistent with known states of affairs. With this as a background I shall now return to the passage cited from Jahoda near the start of this paper. Jahoda, in common with many laymen, possibly some scientists and some philosophers, believes that 'the mind of the observer is a necessary ingredient of the structure of [quantum] theory.' In this context the phrase 'mind of the observer' merits further attention. 'Mentality,' as studied by behaviour scientists, involves only the behaviour of organisms. But behaviour is something entirely physical (as far as its publicly observable aspects are concerned) carried out by physical systems, in this case organisms. The fact that behaviour scientists, such as experimental psychologists and ethologists, more often than not, describe organismic behaviour in terms of a descriptive vocabulary that differs drastically from that used by physicists is due to the level of analysis adopted. This also applies to physics and chemistry, where for many purposes a system, such as a metal, is described in terms of system-specific parameters rather than in terms of more reductive state parameters of the quarks and electrons (etc.) that constitute the atoms of the metal according to elementary particle physics. Yet, the widespread use of 'system-specific' language in all branches of science does not mean that these systems are not composed of matter which is capable of further analyses into subconstituents of the kinds discussed. The fact that scientists acknowledge the possibility of reductionism as something that is possible in principle does not mean that they always demand fully reductionist explanations in practice, since this would neither be profitable nor desirable. Hence, mentality, as far as its behavioural aspects are concerned, need only be related to those organism-specific behaviour descriptors that are deemed useful, without thereby renouncing reductionism, i.e., that the behaviour concerned is thought to be entirely describable - in principle - in terms of physical state parameters, down to the parameters of elementary particles (etc.). In practice, however, the reductionist idea is unattainable, except in elementary particle physics.

When Jahoda refers to the 'mind of the observer' in quantum mechanics, it must be realized that the process of observation is subdivided into activities of one or more observing instruments and human beings who inspect the recordings of the observing instruments. This will be discussed more fully in the following section. According to quantum mechanics the observing system may induce state transitions in the physical system that is being observed. It could be argued that a human being intervenes in the 'observation' by setting and/or pressing appropriate buttons of the observing instrument. Yet, these activities are purely behavioural, and, hence, in principle entirely describable in terms of the physical parameters of the human experimenter, when the latter is considered as a physical system or an automaton. According to materialistic dualism there are no reasons for asserting that the human experimenter's intervention in any experiment, that leads to a change in quantum states of an observed system, is 'caused' by conscious states of the experimenter. Hence, while an adherent of materialistic dualism does not deny an interaction between a human experimenter and observing instrument(s), and a further interaction between the observing instrument and the observed system, he denies that any of these interactions are 'caused' by human consciousness. In fact, one must ask where in any textbooks or expositions of quantum mechanics the term 'consciousness' appears.

Nobody wishes to deny that human beings have transformed the surface of our planet in manifold ways, by building towns and villages, by farming, by creating diverse aspects of the most varied cultures (represented, for example, by printed books, musical scores and so forth) and civilizations. Yet, all these transactions can be viewed as the activities of sufficiently complex systems of matter based on purely physical types of interactions, some of these interactions being those of MSAMCs. Which level of reduction one adopts for describing human behaviour depends on the type of behaviour concerned, and on the systems involved. For instance, in some cases scientists have to discuss human behaviour in relation to macromolecular systems. (Thus, there exist hypotheses and consistent evidence that human brain dopamine receptors, i.e., molecules or macromolecules, are involved in the causation of schizophrenia.[14])

I conclude that Marie Jahoda's views - and those of like-minded people - concerning the foundations of quantum mechanics rest on a misleading interpretation of the act of observation, and that for those who embrace materialistic dualism the consciousness of the human experimenter in quantum mechanical 'observations' is an irrelevancy. Jahoda appears to be a somewhat disguised Cartesian interaction dualist, and interaction dualism is not a necessary metaphysical prerequisite for the foundations of quantum mechanics, since, instead, one can adopt materialistic dualism. In the following section I shall strengthen the preceding argument by referring to Bohm's arguments on the nature of observation in quantum mechanics. His arguments appear to be complementary to mine, and when our arguments are taken in conjunction they lead to an interpretation of observations in relation to quantum mechanics that is fully consistent with materialism and a mechanistic outlook.

V

In view of the preceding discussion it is desirable to probe a little deeper into the quantum theory of the observation process. There exists a very sizeable literature on this subject. I shall only refer to Bohm's chapter relating to this topic in his well-known text. He writes:[15]

> If the quantum theory is to be able to provide a complete description of everything that can [physically] happen in the world, however, it should be able to describe the process of observation itself in terms of the wave functions of the observing apparatus and those of the system under observation. Further, in principle, it ought to be able to describe the human investigator as he looks at the observing apparatus and learns what the results of the experiment are, this time in terms of the wave functions of the various atoms that make up the investigator, as well as those of the observing apparatus and the system under observation. In other words, the quantum theory could not be regarded as a complete logical system unless it contained within it a prescription in principle for how all of these problems were to be dealt with.

Bohm claims that he has demonstrated that 'one can treat these problems within the framework of the quantum theory.' This, of course, is no guarantee that quantum theory is any 'ultimately valid' theory for dealing with the domain of discourse to which it is currently being applied, since no theory can be shown to be true. It certainly cannot be claimed that quantum theory could 'provide a complete description of everything that can [physically] happen in the world', since quantum theory is simply not being applied in explicit form when dealing with certain physical systems. Yet, this point is trivial, and must not detract from Bohm's further remarks.

Bohm assumes, almost from the start, that 'it is correct for us to approximate the relationship between the investigator and his observing apparatus, in terms of the simplified notion that these two are two separate and distinct systems interacting only according to the laws of classical physics.' Bohm further postulates that 'the various possible configurations or states of the measuring apparatus, corresponding to the different results of a measurement, can correctly be regarded as existing completely separately from and independently of all human observers. The quantum theory of the measurement process can in this way be reduced to a description of a relation between the state of the system under investigation and the state of some classically describable part of the observing apparatus.' Very significantly, Bohm adds that 'If a sharp distinction could not be made between the observer and the system observed, scientific research as we know it could not be carried out, because the observer would not know which aspect of an observation originated in himself and which originated in the outside systems of interest.'

As an example Bohm argues that

> If we wish to observe the position and momentum of an electron to a quantum level of accuracy, we must regard the electron, plus the light quanta used in making the observation, as part of an indivisible combined system. Eventually, however, these light quanta are able to activate classically describable processes, e.g. the production of spots on a photographic plate, and from there on the distinction [between observing system and observed system] may be drawn.

Although Bohm takes some trouble to 'consider' the problem of how far into the brain [of the human experimenter involved in an observation] the point of distinction between the observer and what is observed can be pushed,' he concludes - in my opinion correctly - that he wishes 'to stress that the question is completely irrelevant as far as the theory of measurement is concerned since...it is necessary only to carry analysis to some classically describable stage of the apparatus.'

It is, perhaps, tempting to consider those numerous observations, say of stars in our galaxy, that were performed simply by people looking at the stars through a telescope. Nowadays everything that is being 'observed' can in principle - and as a rule in practice - be recorded by means of appropriate recording systems. At any rate, it would be unreasonable to suggest that what a photographic plate or a moving film (etc.) of a camera, attached to an observing system, records is in any way affected by what goes on in the brain of any person who subsequently inspects that plate or film. Hence, when seen from this point of view, neither the human brain that is involved in inspecting a record of an observation, nor possible states of consciousness that may arise in association with the inspection process, alter the records of the observation in any way. To think otherwise, would be like arguing that reading a book changes the printed text in some significant way. Thus, whether man is conscious while inspecting a recorded event of an observation is irrelevant. The same recorded event might be fed into an automaton which is not conscious in the human sense, but which may serve to analyze the record in accordance with an investigation program.

Ultimately, of course, human beings make use of the analyses of observations, irrespective of whether these analyses are obtained in the first instance by observing automata or by observing instruments combined with human observers. This certainly leads to new questions about the utilization of observational information by the cognitive machinery of human brains. This machinery may be instrumental in the formulation of new hypotheses and of new theories formulated in terms of some of these hypotheses. Yet, for the purpose of observation alone, any human being involved in an observation may be regarded as a black box. The problem of how human beings could formulate new hypotheses and new theories in terms of their given brain components is an issue that concerns neuropsychology. If quantum mechanics should, at some future stage, help to explain the performances of brains that are involved in producing various cognitive (etc.) transactions, then this would be very useful, yet, if so, this would not be a consequence of the consciousness of thought processes involved in applying quantum mechanics. This conclusion at least applies to those who embrace materialistic dualism or some other brand of materialism.

VI

Having suggested, in common with many others, that organisms are self-organizing automata (the class of automata concerned comprising automata that are capable of having conscious experiences), I must take some notice of those who have questioned this assumption. Some people have paid, probably quite excessive, attention to Wigner's[16], contention that, to cite Landsberg:[17]

> [Q]uantum mechanics leads to practically zero probability for the existence of self-reproducing states. He [Wigner] concluded that either present quantum mechanics cannot furnish an explanation of all that exists, or it should be modified to include aspects, such as consciousness which are not needed when dealing with purely physical problems.

Although Landsberg, in the passage just cited, argues - in agreement with my previous conclusions - that consciousness is 'not needed when dealing with purely physical problems', this view was not shared by Wigner.[18] He believed that there was an irreducible role of consciousness in the theory of measurement in quantum mechanics. In his opinion the need for introducing consciousness into measurement is called for by the participation of a living system, namely a human observer. In the preceding section I discussed at length my view that - *pace* Wigner, Jahoda and others - as far as the theory of measurement is concerned, quantum mechanics does not have to be supplemented by assumption of the 'consciousness of the observer.'

I shall, however, quote some of Wigner's views that seem to have led him to invoke human consciousness as a *potential* (as distinct from an actual) helpmate for explaining self-reproducing states. (I shall also cite Wigner's footnotes, as notes to the present paper. Some of Wigner's arguments rest conspicuously on his footnotes, and I shall comment on his footnotes not in my text, but in the notes where I quote his footnotes.) Wigner[19] was influenced by Elsasser's[20] opinion (and relevant arguments) that one has to supplement quantum mechanics with *biotonic* laws, not contained in quantum mechanics, when dealing with various phenomena in the life sciences. Whereas Elsasser believed that biotonic laws leave quantum mechanics intact, Wigner argued that 'It is more likely that the present laws and concepts of quantum mechanics will have to undergo modifications before they can be applied to the problems of life.' He continued his argument as follows:

For reasons which are not quite clear, the phenomenon of consciousness has become taboo in scientific discussions. Nevertheless, as one can see, for instance from von Neumann's brilliant discussion of the process of quantum mechanical measurement,[21] even the laws of quantum mechanics cannot be formulated, with all their implications, without recourse to the concept of consciousness.[22] It is very likely that those who deny the reality of consciousness only mean that the external world can be completely described without reference to the consciousness of others, that is, that the motion of matter (in the broadest sense of this word) is not influenced by consciousness, even though consciousness is obviously influenced by the motion of matter. According to the view attributed in the preceding paragraph [of Wigner's paper] to Elsasser, this view is neither correct, nor false, but meaningless because it is, even in principle, impossible to describe the motion of all matters by the laws of physics. In particular, it should be impossible to describe living matter in terms of the laws of physics. It seems more likely, however, that this view is incorrect and that living matter is actually influenced by what it clearly influences: consciousness. The description of this phenomenon clearly needs incorporation into our laws of nature of concepts which are foreign to the present laws of physics. Perhaps the relation of consciousness to matter is not too dissimilar to the relation of light to matter, as it was known in the last century: matter clearly influenced the motion of light but no phenomenon such as the Compton effect was known at that time which would have shown that light can directly influence the motion of matter. Nevertheless, the 'reality' of light was never doubted.

This argument by Wigner tries to justify interaction-dualism by an analogy based on light-matter interaction. Arguments by analogy, however, are notoriously weak. Although to Wigner interaction-dualism may seem more likely than epiphenomenalism (whose position he stated in rough terms, without referring to it by name), the latter, in the guise of dualistic materialism remains *a priori* a possibility, and to its adherents more likely than interaction-dualism. Elsasser's arguments do not make epiphenomenalism meaningless, since Elsasser's belief that physics is inadequate to deal with living systems, and that biotonic laws are needed, remains as metaphysical as interaction-dualism (which Wigner espouses) and neither interaction-dualism nor Elsasser's arguments can be used as *a priori* arguments against the (equally metaphysical) position of dualistic materialism. Wigner is simply stating his metaphysical preferences, whereas I am advocating a strongly opposed metaphysical position. Undoubtedly, anyone who accepts interaction-dualism (e.g., Sir John Eccles[23] or Wigner among many others, including many religious people) will accept that consciousness could have causal effects on brain states, and possibly on other matter states. In particular, Wigner seems to believe that consciousness is likely to play a more universal part in biological phenomena and, judging by the context in which he affirms this belief, he may have self-reproduction in mind, as a conceivable application of his 'augmented' quantum mechanics or basic physics. Significantly, he states near the end of his paper,[24] devoted primarily to self-reproduction, that he 'does not wish to close [his] article without admitting that his firm conviction of the existence of biotonic laws stems from the overwhelming phenomenon of consciousness.' Thus, consciousness, presumably as expressed via (unknown!) biotonic laws, ensures, according to Wigner, jointly with a modified quantum mechanics, the occurrence of self-reproduction (e.g., of DNA). Those who think otherwise, including the present writer, think that there is no need for (in any case unknown) biotonic laws, and that quantum mechanics alone should suffice to explain self-reproduction. This type of metaphysics, taken jointly with the dualistic materialism discussed here and elsewhere seems to me to be more parsimonious than dragging in consciousness (via unknown biotonic laws) as a causative agent.

Whenever people get held up with the applicability of quantum mechanics in relation to living systems, there exists a temptation for some of them to invoke some vitalist agent, such as 'consciousness' (e.g. by resort to interaction-dualism via biotonic laws). Resort to vitalist agents did not originate with Wigner but has a venerable ancestry. Whenever people got seriously stuck with available forms of mechanistic explanations of the structure, functioning, or behaviour of organisms, some of them would give up in despair and resort to metaphorical agents with mystical powers (accompanied often by 'proofs' that no conceivable mechanisms of physics could possibly account for certain biological phenomena). Typical of these agents are McDougall's[25] striving 'horme', Broad's[26] 'psychic factor', H.H. Price's[27] 'psychic ether', Driesch's[28] 'entelechies', Kapp's[29] 'diathetes', Eccles'[30] 'influences', Teilhard de Chardin's[31] 'omega point' and 'noosphere', and Bergson's[32] 'elan vital'. Some of this unashamed resort to vitalism has been summarily dismissed by many others.[33] For example, Huxley[34] remarked that resort to Bergson's 'elan vital' is comparable to 'explaining' the mode of propulsion of a railway train in terms of an 'elan locomotive'. Although consciousness, unlike the other vitalist agents, is known to exist, its use as a vitalistic agent via biotonic laws, as suggested by Wigner, seems as yet, uncalled for in order to explain phenomena such as the self-reproduction of DNA (for reasons stated below).

Landsberg[35] showed that Wigner's contention, concerning the alleged inadequacy of quantum mechanics for explaining self-reproducing states, depends very sensitively on the kinds of assumptions made. With more relaxed, but at least equiplausible, assumptions quantum mechanics seems to be consistent with self-reproducing states. Yet, even Landsberg's approach will sooner or later have to be replaced, if possible, by more specific calculations that take the detailed structure of DNA into account, as Ageno[36] has argued. This, admittedly, may be asking for something that is at present perhaps totally beyond the reach of sufficiently accurate quantum mechanical computations and existing quantum chemical formulations. (Landsberg and Wigner[37] certainly agreed, in reply to Ageno, that 'self-reproduction is one of the properties that has to be discussed from the point of view of quantum mechanics in connection with living objects.') Ageno insisted - in my opinion validly - that one must take into account the specificity of structures that typify living organisms if one wishes to reach any decisive conclusions. The state of a quantum mechanical system can be determined by means of its Hamiltonian operator. Any arguments based on a fairly general, (although statistically constrained) and, hence, unspecified, Hamiltonian operator in the types of arguments advanced by Wigner and Landsberg is unsatisfactory. We can be sure that the Hamiltonian operators, of self-replicating parts of organisms have in-built features that are not obvious in any partially arbitrary type of Hamiltonian operator, and may not even be present in many types of Hamiltonian operators. Hence, Wigner's and Landsberg's resort to general, explicitly unspecified, Hamiltonian operators of quantum mechanical systems is bound to lack explicit reference to those aspects of the structure of self-replicating systems which is likely to be critical in a satisfactory quantum mechanical solution of the problem. It seems, therefore, unwarranted at this stage of the art to argue that quantum mechanics is inadequate for explaining self-replication, and with this, I believe, Landsberg seems to agree.

The difficulty of the problematics can be seen, when one realizes that, when dealing with complex self-replicating macromolecules, such as DNA and the effect on these macromolecules of other macromolecules, such as DNA-dependent DNA polymerases, one simply does not have the mathematical know-how available for applying quantum mechanics to such complex systems. To make wood-and-wire models of DNA molecules, as Crick and Watson did, was a great feat. Yet, this is a long way from specifying in explicit form the appropriate Hamiltonian operator for the atoms and inter-atomic bonds which the wood-and-wire structure represents in grossly simplified form (with complete neglect of the details of the electronic structure of the DNA molecule). Similar remarks apply when we consider the quantum mechanical description of the states and modes of interactions of the DNA-dependent DNA polymerase. I conclude that neither Jahoda's nor Wigner's arguments necessitate a supplementation of quantum mechanics by dualism, and that materialism remains a viable proposition.

Notes

1 Jahoda, M., 'Wholes and Parts, Meaning and Mechanism', *Nature* 296, (1982), 497-498.
2 Wassermann, G.D., 'Reply to Popper's Attack on Epiphenomenalism', *Mind* 88, (1979), 572-575.
3 Wassermann, G.D., 'Materialism and Mentality', *Review of Metaphysics* 35, (1982), 715-729.
4 Wassermann, G.D., 'On the Nature of the Theory of Evolutioni, *Philosophy of Science* 48, (1981), 416-437.
5 Wassermann, G.D., *Brains and Reasoning* (London: Macmillan, 1974) and Wassermann G.D. A Philosophy of Matter and Mind (Aldershot: Ashgate, 1994) Section 1.2.7.
6 Cf. Dirac, P.A.M., *The Principles of Quantum Mechanics* (4th edn.) (Oxford: Clarendon Press, 1958); Heisenberg, W., *The Physical Principles of the Quantum Theory* (Chicago: University of Chicago Press, 1930); Born, M. and Jordan, P., *Elementare Quantenmechanik* (Berlin: Springer, 1930); Kramers, H.A., *Quantum Mechanics* (Amsterdam: North-Holland, 1957); Kemble, E.C., *The Fundamental Principles of Quantum Mechanics* (New

York: McGraw-Hill, 1937); Messiah, A., *QuantumMechanics* (Vols 1 and 2) (Amsterdam: North-Holland, 1965).

7 Turing, A.M., 'On Computable Numbers, With an Application to the Entscheidungsproblem', *Proc. LodonMath. Soc.* Series 2, 47 (1937), 230-265, and *ibid* 43 (1937), 544-546; Turing, A.M. in *Minds and Machines* (ed.) A.R. Anderson (Englewood Cliffs, N.J.: Prentice-Hall, 1950), pp 4ff.

8 See also Wang, H. *A Survey of Mathematical Logic* (Amsterdam: North-Holland, 1963) for a formal definition of a Turing Machine.

9 Cf. Wassermann, G.D., *Molecular Control of Cell Differentiation and Morphogenesis* (New York: Marcel Dekker, 1972); Wassermann, G.D., *NeurobiologicalTheory ofPsychologicalPhenomena* (London: Macmillan, 1978).

10 Op. cit. note 3.

11 Op. cit. note 3, where I have also given extensive discussions and references to other work concerning the mind-body problem.

12 Schweber, S.S. *An Introduction to Relativistic QuantumFieldTheory* (New York: Harper and Row, 1961).

13 For a detailed reference to such theories see the list given in note 3.

14 Lee, T., Seeman, P., Tourtelotte, W.W., Farley, I.J., and Hornykiewicz, O., 'Binding of ^{3}H-neuroleptics and ^{3}H-apomorphine in Schizophrenic Brains.' *Nature* 274 (1978), 897-900; see also Hornykiewicz, O., 'Brain Catecholamines in Schizophrenia - a Good Case for Noradrenaline,' *Nature* 299 (1982), 484-486.

15 Bohm, D., *Quantum theory* (Englewood Cliffs, N.J.: Prentice-Hall, 1951).

16 Wigner, E.P., 'The Probability of the Existence of a Self-Reproducing Unit', in *The Logic of Personal Knowledge, Essays Presented to Michael Polanyi on his Seventieth Birthday* (London: Routledge and Kegan Paul, 1961), pp. 231-238.

17 Landsberg, P.T., 'Does Quantum Mechanics Exclude Life?' *Nature* 203 (1964), 928-930.

18 Wigner, E.P., *Symmetries and Reflections: Scientific Essays* (Bloomington: Indiana University Press, 1975), Chps. 12 & 13; see also Mackinnon, E., Review of *The Interpretation of Quantum Mechanics* by Jeffrey Bub (Dodrecht: D. Reidel, 1974) in *Philosophia* 10 (1981), 89-121, p. 110.

19 Op. cit. note 16, p. 232.

20 Elsasser, W.M., *ThePhysical Foundations ofBiology* (London: Pergamon, 1958).

21 Wigner cites J. von Neumann *Mathematische Grundlagen der Quantenmechanik* (Berlin: Springer, 1932) (English trans., Princeton: Princeton University Press, 1955), Ch. 6, p. 420. Wigner also notes in his

footnote that Heisenberg (*Daedalus* 87 (1958), p. 100)., 'puts it even more concisely and picturesquely: "The conception of objective reality...has thus evaporated...into the transparent clarity of a mathematics that represents no longer behaviour of elementary particles but rather our knowledge of this behaviour." Wigner's footnote to his text, however, seems to imply that 'knowledge' must be of a conscious kind. An advocate of dualistic materialism, like myself, can simply retort that 'knowledge' is physically represented by brain representations (e.g. engrams) and that physical activations of these representations may be accompanied by conscious states, which give an additional (conscious) representation of 'knowledge'. Hence 'knowledge' of the behaviour of elementary particles (etc.) does not demand - *pace* Wigner and others - that the physical brain representations of this knowledge are caused by consciousness.

22 At this point in a footnote to his text, Wigner argues: 'It is interesting from a psychological-epistemological point of view that, although consciousness is the only phenomenon for which we have direct evidence, many people deny its reality. The question "If all that exists are some complicated chemical processes in your brain, why do you care what those processes are?" is countered with evasion. One is led to believe that, as explained in [Wigner's] text, the word 'reality' does not have the same meaning for all of us.' Wigner, in this footnote to his text, does not seem to appreciate that materialists, like myself (cf. op. cit. note 9 (1978)), must try to demonstrate the self-consistency of materialism, i.e. that brain states can account - in principle - for any known *type* of behaviour. Consciousness is not being denied, but is also claimed to be a property of matter (see the text). According to this point of view the formulation of the laws of quantum mechanics was accomplished by changes of brain states that led to other physical states of the originators of quantum mechanics (such as writing down the laws of quantum mechanics).

23 Eccles, J.C., *The Neurophysiological Basis of Mind* (Oxford: Clarendon Press, 1953); Cf. Discussions of Eccles' position by J.J.C. Smart in *Between Science and Philosophy* (New York: Random House, 1963), and by A. Flew in *A Rational Animal* (Oxford: Clarendon Press, 1968), pp. 118-119; see also my footnote in op. cit. note 3 and *ibid.* note 2, for further references to the debate between materialists and interaction-dualists.

24 Op. cit. note 16, p. 237.

25 McDougall, W., *Modern Materialism and Emergent Evolution* (2nd edn.) (London: Methuen, 1934).

26 Broad, C.D., *The Mind and Its Place in Nature* (London: Kegan Paul, Trench, Truber and Co., 1925).

27 Price, H.H., 'Haunting and the 'Psychic Ether' Hypothesis with some preliminary Reflections on the present Condition and possible Future of Psychical Research' *Proc. Soc. for Psychical Research* 45 (1939), 307-343.

28 Driesch, H. *The Science and Philosophy of the Organism* (London: Black, 1929).

29 Kapp, R.O., *Mind, Life and Body* (London: Constable, 1951).

30 Eccles, J.C., op. cit. note 23.

31 Teilhard de Chardin, P., *The Phenomenon of Man* (London: Collins, 1960).

32 Bergson, H., *Creative Evolution* (London: Macmillan, 1914).

33 Cf. Medawar, P.B., Review of the *Phenomenon of Man, Mind* 70 (1961), 99-106; Wassermann, G.D., op. cit. note 5, section 4.3A.

34 Huxley, J.S., *Evolution: the Modern Synthesis* (London: Allen and Unwin), 1942), p. 458.

35 Op. cit. note 17.

36 Ageno, M., 'Does Quantum Mechanics Exclude Life?' *Nature* 205 (1965), 1306-1307.

37 Wigner, E.P., and Landsberg, P.T., *Nature* 205 (1965), 1307.

PS. This paper was previously published in *Nature and System* (1983) 5, 3-19. (The paper was also refereed for the journal *Erkenntnis* who were also willing to accept it. But it had by then already been accepted by *Nature and System*.)

20 Some comments on methods and statements in parapsychology and other sciences*

20.1 Introduction

During the last twenty years experimental parapsychology has made great strides. The development of experimental methods peculiar to that subject have raised numerous general methodological and philosophical questions, which have not all been adequately discussed. Sceptics found fault with the details of earlier experiments. Gradually experiments were so much improved and reprehensible errors of methodology so thoroughly stamped out that the sceptics found no more justifiable fault in the experimental set up.[1] Criticisms began therefore to shift to philosophical issues, which primarily concerned the *interpretation* of the results obtained. Three of the most important of these issues are: (1) whether parapsychological experiments are repeatable and in which sense they are repeatable, and also whether specific effects are predictable; (2) whether adequate control methods exist; (3) whether statistics is on logical grounds a reliable tool for assessing the significance of experimental data in this field.

Recently G. Spencer Brown has devoted his attention to these issues and in fact has challenged current interpretations of parapsychological experiments.[2] He has

---

* Received 4.viii.54. This paper is based on numerous discussions of the writer with a colleague, Dr R. Harrop, to whom some of the leading ideas and some of the details are due. It was not found possible to present our partly divergent views satisfactorily in a joint paper, and it was therefore decided that one of us should write a paper presenting his own point of view. Many of the arguments I give do, however, owe much to the discussions I have had with Dr Harrop [who became a professor at Simon Fraser University Canada (added in 1996)]

argued that in parapsychology, experiments are neither repeatable nor repeatable *at will*; that there exist no adequate control series; and finally that the data of parapsychology only throw doubts on the applicability of significance tests, and do not support customary interpretations in terms of telepathy, clairvoyance, and similar notions. It seems that some of his propositions, if accepted, could, by slight generalization, invalidate many currently accepted demonstrations of phenomena in biology and psychology. For these very reasons much of the present discussion may be of concern to biologists, psychologists, and other scientists, and not only to parapsychologists. In fact, the extensive discussion,[3] which has recently arisen from Brown's arguments is believed to be of general interest and the present paper is intended as a contribution to it. Some of my conclusions may appear 'obvious'. Yet Brown's article in *Nature*[4] and his broadcast[5] suggest that the present comments are not superfluous.

I primarily wish to analyse the *logical basis* of Brown's statements. However, this discussion does not pretend to be exhaustive. Certainly, a general treatment of the topics touched on would require a book rather than a paper. Neither is there any intention of going into any details or discussions of experimental parapsychology that are not relevant to the logic of the issue.

My aim is to distinguish different *types of statements* made in psychology, biology, and parapsychology and to show that this distinction is essential when arguing about repeatability and predictability. Brown's views on repeatability will be repudiated because in the writer's opinion he has failed to make this distinction. After discussing the nature of control experiments very briefly, and unmasking an apparently erroneous assumption in Brown's arguments, I shall discuss the validity of statistics when applied to psychical research. Again, I have no ambition to discuss the foundations of significance tests. It is only my aim to suggest that the formalism of statistics can be reinterpreted. When this is done the usual interpretations based on the significance tests are seen to be not in the least invalidated by Brown's arguments. On the contrary, it will be pointed out that lack of adequate randomisation and similar effects suggested by Brown could not plausibly account for the known facts of parapsychology.

20.2 Repeatability and predictability

It is not easy to give a general definition of the concept of repeatability. In experiments in which *measurements* are performed repeatability may be defined as follows. If similar sets of measurements, which were carried out under standardised conditions, are compared and it is found that the results of the measurements for each set fall within permitted (arbitrarily assigned) limits (e.g.

statistical limits of confidence), then the results may be said to be repeatable. There are other situations also where repeatability of a phenomenon has apparently occurred but where no measurements were made. For example, Sir Wilder Penfield and others have electrically stimulated similar regions of the cerebral cortex of many of their patients, frequently obtaining similar responses. Class characteristics may in this case be used to delimit what we call a repeatable phenomenon. It is, of course, important to realize that irrespective of whether measurements can or cannot be performed, the permitted limits of the classes which determine the repeatable phenomena are always dependent on arbitrary conventions. Moreover, within these limits variations of the class members are permitted.

Evidently, according to this definition of repeatability, a phenomenon, even if it occurred only twice, was repeatable. Scientists, however, are not just interested in whether a phenomenon occurred more than once, but whether, when sufficient repeat-experiments were carried out the phenomenon occurred *so frequently* as to conform to prescribed standards of significance. In future when I refer to the repeatability of a phenomenon I shall invariably imply that the phenomenon was *significantly repeatable,* i.e. that the frequency of its occurrence meets prescribed standards of significance.

Science is ultimately devoted to the discovery of relations or classes of such relations among phenomena, or to classes of *similar* events or phenomena, and without repeatability such relations would not seem possible. Nevertheless, the need for repeatability does not imply that there is only one type of scientific statement that refers to repeatable phenomena.

It seems to me essential to distinguish at least *two types* of statements which refer to the repeatability of scientific phenomena. Often scientists only state that there *was* repeatability of a phenomenon for long periods of time, but no prediction is made that there *will be* future repeatability of this phenomenon. Such statements may be *called historical-repetition statements.* For example, palaeontologists may state that repeatedly, generation after generation, there occurred dinosaurs which ultimately became extinct. This is an inference from repeated observations on fossil finds. However, few people would like to predict that living dinosaurs will be found next year in the streets of London. In fact, there is no logical need for any form of prediction in order to make the statement about the past creditable. The recognition that dinosaurs existed is based on *agreed* standards of significance, which in this case are not statistical but expressed by a majority opinion of palaeontologists. Similarly biologists may point out that certain bacterial cultures showed in the *past* no resistance to penicillin for long periods of time, during many generations, until suddenly mutant strains evolved, and thus an irreversible change occurred.

It is obvious that historical-repetition statements require no predictability of phenomena, since, by definition, they refer exclusively to the past and never to the future. These statements are therefore usually not of pragmatic value when applying science to forecasts of future situations; but they may be of great indirect value when formulating theories (e.g. a theory of evolution). For example, it is certainly important for medical men to know that drug resistance may develop in mutant strains of bacteria, even if they cannot predict whether in an individual case such drug resistance will occur.

Perhaps the most important characteristic of historical-repetition statements is that they invariably refer to a class or classes which comprise only a finite number of elements. These statements therefore contain typical terms such as *some or sometimes.* To illustrate this we consider a few typical examples. Zoologists claim that *some* calves, with five legs *have been* born. Yet no claim is made in this case that such a calf will ever be born again or that all calves will have five legs. Similarly, parapsychologists only claim that *some subjects sometimes have* shown for long periods of time an unusual guessing behaviour in card guessing experiments, which was significant by certain standards of significance. However, no claim is made that *every* subject when tested under similar conditions *will always* show such behaviour.

Psychologists, biologists, and parapsychologists deal with organisms which do not show a *static type of ordered behaviour.* A certain behaviour pattern of an organism may be elicited with proper stimuli repeatedly during long periods of time (so as to be significant) and then, often quite unpredictably and sometimes suddenly, the usual stimulus may fail to elicit the familiar pattern. The decline of I.Q. with age above a certain age for the average of the population and the corresponding decline of brain weight are other examples of the change of human abilities and organismic properties with time. Musicians or mathematicians of genius, whose powers of expression suddenly or gradually declined are known, and the subjects in parapsychology experiments whose scoring ability peters out after thousands of successful runs are examples of the same type. We can always assert in such cases what *has* happened, or what faculties did exist, but we can rarely, with any reasonable confidence, predict what will happen. Historical repetition statements are therefore most appropriate to such situations.

The second type of statements which I shall consider are *generalisations.* Generalisations are statements which refer (usually) to a class or to classes which have an *infinite* number of *hypothetical elements* (e.g. infinite populations in statistics or possible values of continuous variables in physics). Generalisations are supported by a finite number of observed data and make *inductions* from these to an infinite set of hypothetical data. The 'exact' laws of science (e.g. Ohm's law) are typical examples. Again, all probability statements are inductive. It is, of course, well known that generalisations have no predictive *certainty* and they only represent attempts to make reasonable posits about future events.

The contrast between historical-repetition statements and generalisations is striking. The former demand no predictability of phenomena and attempt no prediction, while the latter do. The former refer to a finite class whose elements represent *actual* observations, while the latter refer to an infinite class of hypothetical elements. Moreover, historical repetition statements refer only to the past, whereas inductions refer both to past and future. Perhaps most important of all, historical-repetition statements may refer to a dynamic state of order which permits sudden, irreversible, unpredictable, changes (e.g. new types of mutations), whereas generalisations require a well defined future behaviour of systems. Of course the preceding distinction does not exclude that an historical-repetition statements may contain a limited amount of generalisation. It may state, for example, that all members of the finite class which *were* made the subject of a past observation showed a particular characteristic class feature. For example, *all* the dinosaurs *which were found* had a skeleton. However, the degree of generalisation of historical-repetition statements differs from generalisation statements in that it never includes such phrases as *must always or will always or tends always to show.*

Generalisation statements may only be used in biology, psychology, and parapsychology provided there are weighty reasons to assume that for a future time period, specified in the statement, certain predicted regularities will not change. Life insurance, for example, is based on inductive posits. Yet, even insurance companies may have to change their posits from time to time if, owing to *unforeseeable* causes, mortality rates should change. Thus there may be inductions with changing posits, which involve predictabilities limited to the period during which each posit applies.

In the light of the foregoing remarks we can now examine Brown's statements on repeatability and predictability in parapsychology. These are: (1) that, as a rule, identical experimental conditions do not bring about a repetition of results;[6] (2) that the results are not repeatable *at will*;[7] (3) that 'to demonstrate a causal tendency we must be able to say beforehand what sort of results we are going to get'.[8] He complains, for example, that in card guessing experiments '...nobody can predict the coming of a successful guess and nobody knows when one has been made until the results are checked up afterwards'.[9]

It is seen that Brown's statements depend on his definition of the concept of 'demonstrability' of phenomena. If we were to agree with Brown that only those phenomena are to be termed demonstrable which are repeatable *at will*, or whose causal tendency may be predicted beforehand, then certainly, parapsychological phenomena, and many other phenomena of science, would not be demonstrable.

It must be realised that Brown seems to *define* only those phenomena as demonstrable, which permit *generalisation statements.* Thus by definition he automatically excludes all those phenomena from being demonstrable which require historical-repetition statements.

Parapsychologists make essentially historical-repetition statements, which refer to the sort of causal tendencies that *were* observed. They do not therefore require statements as to what causal tendencies they are going to get in future experiments (generalisations). Secondly, parapsychological statements, being non-inductive, do not require indefinite repeatability or repeatability *at will.* However, parapsychologists can certainly point to extensive (significant) *past* repeatability in many experimental series, in the sense of historical-repetition statements. For example, Soal[10] found, with two remarkably gifted subjects, repeatability of guessing behaviour, typical for each subject, over long periods of time, in batches of 200 guesses in many thousands of trials. The guessing habits of his subjects exhibited a striking fine-structure of the spacing of the coincidence scores and this was found to be a repeatable phenomenon. Brown's statement that identical experimental conditions do not 'as a rule' bring about a repetition of results is therefore repudiated by this important series of results and similar other extensive series conducted by other workers.[11] These series showed consistent behaviour during tens of thousands of guesses and even in some cases permitted successful limited inductive posits (comparable to those of life-insurance companies). In much of this work people set about purposely to demonstrate the presence of parapsychological ability and did succeed. Here the facts again are at variance with Brown's statements that such attempts to demonstrate 'telepathy', etc., 'nearly always resulted in failure'. In one case, in fact, very striking repeatability occurred. Soal found that one of his subjects showed even after a lapse of several years repeatability of idiosyncratic guessing behaviour, when the same agent was used.[12] Of course, neither Soal nor anyone else ever claimed that there was *detailed* repeatability, guess by guess, nor is it claimed that every single run showed always the fine structure typical of the subject's scoring habit. Nevertheless, the typical fine structure recurred very strikingly and significantly (by usual standards) when similar experimental conditions prevailed. If we only recognised experimental series, which showed guess by guess 'exact' scoring level repeatability, when compared, then certainly we should have to exclude parapsychological phenomena from recognition. However, in that case we should also have to exclude all those parts of science which show experimental series capable of statistical fluctuation. This, I believe, would probably exclude the whole of science. Even physical repeat measurements hardly ever lead to exactly the same experimental figures.

Instead of comparing *individual* guesses, parapsychologists only examine critical ratios for a *whole batch* of guesses and try to see what sort of correspondences occur most frequently batch by batch. Their statements on repeatability therefore always refer to whole batches, and not to single guesses. Even for whole batches fluctuations are permitted and it is not claimed that repeatability means that every batch shows *identical* critical ratios. Instead the ratios are permitted to vary between permitted (arbitrarily assigned) limits.

It is crucial to realise that Soal only claimed that his two subjects showed their remarkable guessing behaviour in the *past* (historical-repetition statement). He did not claim that other experimenters under similar conditions *will always* find similar results, nor that any arbitrary subjects or agents would be suitable, nor that his subjects could continue their feats indefinitely. As his statements are not generalisations there is no logical necessity for such a claim to be upheld.

Brown's insistence that phenomena should be demonstrable *at will* is again another expression of his definition of 'demonstrability' so as to exclude historical-repetition statements from being demonstrable. It is not difficult to give examples which illuminate the inadequacy of his criterion of 'demonstrability' *at will.* Cosmic ray experiments provide an immediate instance. When physicists examine photographic plates which have been exposed to cosmic rays at high altitudes, *they sometimes,* though only rarely, find a particular sort of track which they call a meson track. If a sceptic were to ask a physicist to get him, in his presence, a meson track of a particular kind by exposing a batch of 100 plates for a given length of time at a high altitude, the physicist might be very embarrassed. The whole batch might not show a single track, so that the phenomenon is not demonstrable *at will.* But this hardly means that meson tracks have not been demonstrated. Earthquakes and meteorites in the atmosphere also cannot be demonstrated at will, but few people would deny that they *have been* demonstrated. Again, a competent psychiatrist who may be an excellent hypnotist, may fail, when ordered to hypnotise a randomly chosen subject *at will.* Nobody would suggest nowadays that therefore hypnotism has not been demonstrated. People would only argue that hypnotism just does not work in such a fashion that results can necessarily be produced at will, and they would benignly smile at anyone who made such a demand. Yet only about 100 years ago medical opinion refused to recognise the existence of hypnotism on such grounds, and shortly before that people who accepted the existence of meteorites were made the laughing stock of society. It was then also held that such phenomena could not possibly exist, since they were not repeatable at will. It is seen that the fatuous demand (often made) that parapsychological subjects, if they have any gifts at all, should be able to predict the results of the stock exchange[13] is also based on the assumption that we can prescribe to natural faculties how and when they ought to work. A subject who is good at card guessing, even if he scores very highly, gives a poor performance compared to the feat required in predicting accurately a complicated series of events (e.g. the stock exchange).

Again, Brown's complaint 'that nobody can predict the coming of a successful guess and nobody knows when one has been made until the results have been checked up afterwards' can now be compared with the statement that 'nobody who exposes plates to cosmic rays can predict the coming of a successful meson track exposure and nobody knows when one has occurred until the plates are checked up afterwards'. In other words, in order to demonstrate that a phenomenon *has* happened we do not have to have predicted beforehand *when* it should have happened, nor have we to have predicted the detailed structure of the phenomenon.

20.3 Objectivity

Scientists (particularly psychologists) have spent much effort to make experiments 'objective'. The doctrine of 'objectivity' demands that, given a specified experimental set up, any experimenter should be able to obtain similar results under standard conditions and, if desired, in the presence of witnesses. This stipulation has held a dominant place in modern science.

Parapsychological experiments have revealed[14] a situation which from this point of view is of particular philosophical interest. It was found that the personality structure and attitude of *all* those who partake actively or *passively* in such experiments could be of great importance in deciding the outcome.[15] Thus the personality structure of experimenters, agents (i.e. the person who looked at the cards) subject and possible witnesses present could all be of decisive influence. For example, a change of agents in Soal's experiments could result in an immediate decline of successful scoring and, usually, resubstitution of the same agent would lead to resumption of the previous success. Even the detailed structure of the guesses depended on the personality of the agents used and changed for a given subject with the agents in a consistent manner.[16] Again, Soal found that only two out of 160 subjects tested by him could score persistently significantly above expectation, and these subjects scored in a manner depending on their personality. Other experiments revealed that a change of *experimenters* with the same subjects could change the outcome of the experiments, even if the experimenter played outwardly a passive part, by watching the experiments only. Some experimenters, with gifted subjects, produced consistently successful scores, while others were equally consistently unsuccessful with the same subjects (e.g. Soal and Rhine both experimenting with Mrs E. Garrett using the same technique failed and succeeded respectively). Finally the extensive work on the correlation of personality structure (determined by psychological tests) and scoring ability and scoring habits in parapsychological experiments have revealed notable results.[17] Gardner Murphy has stressed that work in this field has to be 'individualised' because of these findings. The usual 'objective' method is unsuitable, since the personality structure of the experimenter and others can influence the results. We have here what Murphy and others[18] have termed field phenomena, namely a complex interaction between a multiplicity of personalities in a single experiment. It is probably only in the social sciences that we find situations of comparable complexity.

Any statement that repeatability has occurred in parapsychological experiments must therefore take this complex situation into account. It must be stated that conditions were similar, namely, that the same agents, experimenters, witnesses, and subjects were used, that the subject's state of health was similar, etc. Otherwise there is no justification for claiming repeatability, or even to object that there was no repeatability. It appears that Brown's statements, and similar accusations raised by others, that *as a rule* there is no repetition in such experiments, have overlooked the complexity of the situation. Brown has not qualified his statements by reference to particular experiments and without such qualifications his statements become meaningless, since many experiments were performed under different conditions.

20.4 Controls

It has been argued by Spencer Brown and others[19] that parapsychological experiments lack 'satisfactory control series'. He maintains that in psychical research, results deviating significantly from expectation occur in control experiments when experimenters are intent on preventing their occurrence and that they cannot therefore be excluded to order.[20] He particularly emphasises that 'control series in this work cannot be relied upon to give results which are any less curious and 'significant' than the main series'. This suggests a demand that in control experiments the phenomenon to be investigated must never occur and hence that there must be no 'significant' results in the control series. Brown apparently believes that experiments which do not satisfy this requisite are of little value.

A simple example will show that this demand is neither logically required nor practicable. A recent investigation seems to suggest that cigarette smokers are more prone to develop lung cancer than non-smokers. Cigarette smokers thus formed one series (the main series) and non-smokers another series (the control series). However, even non-smokers will in many if not in all cases have been exposed to cigarette smoke at some time. Hence it cannot be claimed that their lungs could not possibly have been in (significant) contact with the substances associated with cigarette smoke. It can only be asserted that cigarette smokers and non-smokers were generally exposed to different concentrations of smoke. Thus, in order to assess the effect of some suspected causal factor, such as cigarette smoke, it is logically not required that the factor be present in one series (the main series) *and completely* absent in another series (the control series). In fact, in many if not in all experiments it is impossible to make sure that a given factor X whose suspected effect is investigated could have no significant effect in the control series or is completely absent from the control series. Scientific control series have apparently one main purpose, namely to establish a *comparison* between two (or more) series of experiments of which one may be the main series and the other the comparison series. Usually a comparison is made either of results obtained from similar systems under changed experimental conditions or of results obtained for different systems under similar conditions. All that matters, if a causal factor is to be demonstrated, is that (two or more) different comparison series show a repeated consistent (significant) *differential behaviour* under different specified experimental conditions. However, there is no need for one of the comparison series to show nothing 'significant' as far as the suspected factor is concerned. For example, if for a gas the volume is measured against pressure at constant temperature and a curve is plotted, and curves are obtained for different temperature values (one curve for each value of temperature) then the different curves may be taken as comparison series. Here the same causal factor (temperature) was present in each comparison series of experiments and affected the measurements, though each curve was affected *differently.* Thus all comparison series (control series, main series) may be permitted to show *significant* influence of the same supposed factor X (e.g. temperature) provided only that this factor consistently affects the different comparison series in a distinctly different manner.

Brown's complaint that significant results in control series have occurred in parapsychological experiments becomes now irrelevant. First of all, in many types of experiments when guesses of subject were matched against shuffled packs of cards for which they were not intended and the same guesses were also matched against the cards for which they were intended *persistent and consistent differential behaviour* in the coincidence scores was found.[21] Secondly, in other experiments (e.g. Soal's) where several significant comparison series emerged each comparison series was consistently distinguishable from any of the others[22] (just as the gas volume curves above for different temperatures). In fact in all cases where this occurred feasible theoretical explanations have been suggested, although these are not important from the present point of view.

20.5 Statistical significance

We come now to a point of major interest. The following discussion is partly of general interest and partly arises from an interesting logical situation which seems unique in parapsychology and which does not seem to arise elsewhere in science. Brown contends[23] that the evidence collected by parapsychologists looks suspiciously like observations which would render the formulae for statistical significance unreliable. His argument may be summarised as follows. He considers statistics as a *model,* which is derived from a set of postulates. One of these is the postulate of the random occurrence of certain events (which occur as specific values of a postulated 'random variable'). For example, parapsychologists usually assume that the cards looked at by the agent are 'randomised'. The postulate of 'randomness' encounters the difficulty that it is not possible for a finite set of 'random numbers' to give a test which could uniquely check them for 'randomness'. The usual empirical tests can only make sure that *specific types of orders* do not occur. Suppose now the empirical process of 'randomisation' used by experimenters, was - in the absence of suitable tests - not as good as is sometimes believed. This, Brown argues, might have serious consequences. First of all, it could manifest itself in a pathological behaviour of tables of random numbers, such as those constructed by Kendal and Babington-Smith. The usual tests for 'randomness' would not have discovered these abnormalities. As biologists, psychologists, and parapsychologists sometimes use such tables this might have serious repercussions. In fact, Brown has hinted that pairing of independent columns of such tables might lead to more significant coincidences than statistics would lead us to expect if the tables were really 'random' and that the structure of such coincidences might in fact resemble the structure of the results obtained in parapsychology.

It is well known that many people felt very impressed by this possibility and thought that Brown had at last 'explained away' parapsychology. However, to check Brown's suggestion Oram[24] has recently paired 100,000 random numbers taken from the tables mentioned. Many of Brown's claims were not confirmed and the results behaved just as the statistical model would lead us to expect, except for a significant decline effect.[25]

The attack on the random tables is only the first of Brown's logical arguments. He suggests further that in parapsychological experiments the means of randomisation might have been influenced by the very processes which parapsychologists claim to have discovered. For example, they claim to be able by mental processes alone to influence the fall of dice and to be able to control by thought the motion of micro-organisms.[26] This possible influence on the empirical means of randomisation, or any other cause which might lead to inadequate randomisation, might then conceivably invalidate the applicability of the postulate of randomness and thus liquidate statistics as a useful model in assessing the significance of parapsychological data. The way in which such a breakdown of statistics could manifest itself would be by the occurrence of certain significant deviations from expectation values, precisely of the type found in parapsychology. Thus, according to Brown, the data of parapsychology need not necessarily suggest telepathy or precognition, etc., but could simply mean that a breakdown of the applicability of the statistical *model* has occurred.

Logically this is a very interesting suggestion and seems, for the moment, to have captured the imagination of many statisticians, particularly those unfamiliar with the facts of parapsychology. Similarly, for some psychologists Brown's arguments opened a splendid opportunity to deny wholesale without further scruples facts which were inconvenient, since they apparently did not agree with accepted theoretical models. However, it was less readily realised that Brown's arguments could in principle raise serious doubts against the use of statistics in the interpretation of facts in any field of research (including psychology) which employs empirical means of randomisation.

Brown's argument in fact depends on the crucial assumption that statistics, as used in practice, *must be considered as a model,* and hence depends on its postulates. I wish, however, to challenge this assumption, which I believe to be superfluous, although it has tacitly been taken for granted by most statisticians. It is suggested that in parapsychology and in many other fields of research statistics need not necessarily be considered as a model.[27] Instead in certain applications statistics can simply be regarded as the source of convenient *rules of calculation* which enable certain data to be expressed in 'normal' form for the purpose of *comparison.* For example, in the application of statistics to parapsychology and other researches statisticians often calculate so-called critical ratios. To compute these they make use of standardised formulae[28] which may be considered purely as *algebraic expressions* and as such *in no way* depend on the postulate of randomisation. The validity of the use of these formulae does not depend on whether the means of randomisation (e.g. tables of random numbers, or chips drawn blindly from a bag) are 'perfect' or not. Nor does it matter for the purpose of computation by what logical arguments these formulae arose in statistics. Now, what really matters in parapsychology and other sciences is *the comparison* of critical ratios for different comparison series of equal size. In other words, the *relative* magnitude of critical ratios for comparison series but not the absolute value of the ratios for each series is the crux of the problem. It is, of course, true that many scientists have posited their claims on the absolute magnitude of critical ratios. Although this may be perfectly justifiable in individual cases, it is of little interest to the present discussion. I wish to stress that there is *no necessity* for relying on the use of absolute values which require the use of statistics as a model. It seems that most, if not all experiments can be interpreted satisfactorily by studying only the *differential behaviour* of critical ratios (or of other suitable 'significance numbers') for different comparison series. Brown's major contention that in parapsychology the axiomatics of statistics might be infringed becomes now irrelevant, since the *comparison* of critical ratios does not in the least require satisfaction of the axioms. Nor is it therefore of interest from the present point of view whether the subject can or cannot influence the means of randomisation. However, with the present method of interpreting statistical calculations, it could still be possible that the method of computation causes the particular structure of the data, or that a preferential guessing pattern of the subject might favour the now permitted non-randomness of the target cards and thus produce the result. These possibilities will now be briefly examined. Both of them will be shown to be apparently incapable, either singly or in combination, to account for the data of parapsychology. It follows then that the new interpretation permits to re-establish the status of the customary parapsychological interpretations to that which they held prior to Brown's criticisms. In other words, *Brown's criticisms are insufficient to reject current interpretations.* This, of course, does not mean that I accept these interpretations necessarily as the best possible. It is also noteworthy that the present method of considering only the relative magnitudes of significance numbers avoids Brown's charge that parapsychologists are guilty of a policy 'whereby when there is reason to believe other causal agents have been eliminated, all statistically significant results are taken as evidence for E.S.P., and only non-significant results are cited as relevant for the applicability of the probability calculus'.

These accusations are only justified as long as statistics is considered as a model. The mere applications of rules of calculation requires no general justification of 'applicability', and thus Brown's arguments no longer apply to this interpretation.

First of all let us suppose that lack of randomisation of target cards, combined with a preferential guessing pattern of the subject had caused the results of most experiments. Many experimenters found that some of the subjects tested by them persistently failed to score successfully, while only a few subjects under similar circumstances scored successfully for long periods of time in a *consistent and persistent* manner. It was found when the guesses of the successful and unsuccessful scorers were both compared with cards randomised by the same process as the target cards, but *for which the guesses were not intended*, that only the successful subjects showed a markedly higher score on the targets than on the not 'intended' cards. It is difficult to see how lack of randomisation could have caused this differential behaviour; the same lack of randomisation should be present in *both* comparison series, since the same methods of randomisation were used. It seems rather that the 'intention' was instrumental in the occurrence of the differential behaviour. An alternative, of course, would be to assume that the subject affected the means of randomisation in such a way as to cause the coincidences. Since the subject was invariably separated from the means of randomisation, this would require some process of the type suggested by parapsychologists, namely, a long distance correlation.

In Soal's experiments in particular the relative magnitude of the critical ratios of various comparison series used changed consistently with change of experimental conditions. Thus, a change of the speed of guessing or a change of agents was regularly and repeatedly accompanied by characteristic changes in the relative magnitude of critical ratios. Most significantly the level of success dropped completely to 'insignificance' when the agents only touched the backs of cards instead of looking at them.[29] If we assume that the guessers always *synchronised* their specific guessing pattern so as to suit the non-random order of the targets, this leads again to difficulties. First of all the guessing pattern would have to adjust itself *consistently* to the change of experimental conditions. But in this case it would again be queer that the scoring of guesses against card orders for which they were not intended and obtained by the same methods never yielded anything significant. This suggests that a preferential guessing pattern could hardly be the correct explanation. Moreover, the fact that the experiments only succeeded when the agents looked at the cards and the fact that the structure of the guess coincidence varied sensitively with the agent used, suggests strongly that there existed a *causal* relation between subject and agent. It is this causal relation which parapsychologists have called E.S.P. Of course, this leaves open the question whether this causal process is reducible to any of the types of biological or physical processes hitherto known to science.

It is also most implausible to assume that the possible synchronisation between the subject's biased guessing behaviour and the non-random order of the cards could produce such *consistent and persistent* results as occurred in the work of Soal[30] and the work of Martin and Stribic[31] and also of Pratt and Woodruff[32] to quote but a few examples. The synchronisation would have to show always deviations of similar sign, since, for example, in the work of Soal and Goldney[33] the deviations for each batch were always positive for the most marked series of coincidences. This would mean that the bias of the non-random order of the target cards would have to favour strongly one direction of deviation. This assumption would be incompatible with the fact that some of the comparison series achieved with the same methods of randomisation led equally consistently and persistently to smaller deviations. If, on the other hand, the synchronisations were only a temporary chance effect, which led sometimes to a positive deviation, then some later work should sometimes show negative deviations. However, the persistence of one-sided deviations in Soal's work, week after week, suggests that this was certainly not the case.

It is also most implausible that the methods of computation should have given rise to any of these persistent effects. The method of computation (i.e. algebrical formulae) could hardly distinguish between different comparison series in a consistent and persistent manner.

Finally, the correlations between personality structure and scoring habits and between the subject's attitude and his scoring habits, which were discovered persistently and consistently in the experiments of Humphrey,[34] Schmeidler,[35] and others could in no way be explained as being due to a lack of randomisation. For example, when the subjects were divided according to two attitudes *A* and *B* into two groups, *A* as a group would score positively above expectation and group *B* negatively below, showing thus a persistent and consistent differential behaviour of sign, provided the same experimenter was used. In these experiments all subjects of both groups directed their guesses simultaneously towards the *same* targets. Moreover, the subjects did not know what method of randomisation was used, so that there could be no learning process involved. Other experiments showed that psychological factors, such as a change of stimulus size or of 'interest' could produce a repeated and immediate correlated response in the subject's scoring.[36] These evidently *psychological* relations suggest strongly that we are here confronted with genuine psychological processes, which though at present no better understood than hypnotism, represent something very different from a statistical freak. In these experiments the faulty means of randomisation would have to be so biased as to favour always the appropriate personality types, which seems most implausible. There are also many people, who argue, probably with justification (e.g. Soal and Thouless) that there are good reasons why statistics is even applicable as a model in parapsychology. In particular they point to the existence of very large critical ratios, which they believe could not be due to a lack of randomisation on account of the null results in cross checks. While I do not deny the importance of their conclusions, I stress that the present considerations are independent of theirs.[37] [Added in 1996. It has been known since 1978 that some of Soal's most prominent results are due to data manipulation and, hence, worthless.]

Notes

[*Note added in proof.* Since this paper was written an important paper was published by G.L. Mangan (*J. Parapsychol.*, 1954, **18**, 209). He obtained a highly significant differential behaviour in psycho-kinesis experiments with dice; the results changing consistently with a change of intention of the experimenter. Like the Soal (etc.) experiments this work does not seem to be explicable in terms of Brown's hypotheses.]

1 Soal, S.G. and Bateman, F., *Modern Experiments in Telepathy*, London 1954.
2 Spencer Brown, G., 'Statistical Significance in Psychial Research', *Nature*, 1953, **172**, 154.
3 Soal and Bateman, op. cit.note 1; Spencer Brown, op. cit; Report to the First International Conference of Parapsychological Studies, Utrecht, 1953; Letter to *Nature*, 1953, **172**, 594; Letter to *J. Soc. Psych. Res.*, 1953, **37**, 210; 'Telepathy and the Laws of Chance', B.B.C. Broadcast, 17th February 1954; S.G. Soal, Report to the First International Congress of Parapsychological Studies, Utrecht, 1953; Letter to *J. Soc. Psych. Res.*, 1954, **37**, 246; S.G. Soal and J.M.F. Stratton and R.H. Thouless, Letter to *Nature*, 1953, **172**, 594.
4 Spencer Brown, G., Op. cit. note 2.
5 Spencer Brown, G., B.B.C. Broadcast, cited in note 3.
6 Spencer Brown, G., Op. cit. note 2
7 Spencer Brown, G., B.B.C. Broadcast, cit. in note 3.
8 Spencer Brown, G., 'Letter to *J. Soc. Psych. Res.*, 1953, **37**, 210.
9 Spencer Brown, G., Op. cit. note 2 and BBC Broadcast cited in note 3.
10 Soal and Bateman, op. cit.note 1; S.G. Soal, 'The Experimental Situation in Psychical Research', The Ninth Frederic W.H. Myers Memorial Lecture, London, 1947.
11 Martin, D.R., and Stribic, F.P., 'Studies in Extra-Sensory Perception (I): An Analysis of 25,000 Trials', *J. Parapsychology*, 1938, **2**, 23; 'Studies in Extra-Sensory Perception (II): A Review of All University of Colorado Experiments'; same, 1940, **4**, 23; B.M. Humphrey, 'The Relation of Some Personality Ratings to E.S.P. Scores', *J. Soc. Psych. Res.*, 1951, **36**, 453; G. Schmeidler, 'E.S.P. Performance and the Rorschach Test', *J. Soc. Psych. Res.*, 1950, **35**, 323.
12 Soal, S.G., Letter fo *J. Soc. Psych. Res.*, 1954, **37**, 246. (This is no longer valid. G.D.W. in proof 1996.)
13 This demand was made by a speaker at the discussion on parapsychology at the British Association in Oxford in 1954 and was widely applauded.

14 Murphy, G., 'Psychical Research and Personality', *Proc. Soc. Psych. Res.*, 1949, **49**, 1.
15 Humphrey, B.M., 'The Relation of Some Personality Ratings to E.S.P. Scores', *J. Soc. Psych. Res.*, 1951, **36**, 453; G. Schmeidler, 'E.S.P. Performance and the Rorschach Test', *J.Soc. Psych. Res.*, 1950, **35**, 323; J. Fraser Nicol and B.M. Humphrey, 'The Exploration of E.S.P. and Human Personality', *J. Amer. Soc. Psych. Res.*, 1953, **47**, 133.
16 Soal and Bateman, op. cit. note 1
17 Humphrey, op. cit. note 1; Schmeidler, op. cit.note 15; Fraser Nicol and Humphrey, op. cit. note 15.
18 Murphy, op. cit. note 14
19 Brown, op. cit. notes 2 and 3.
20 Brown, Broadcast, as cited note 3.
21 Soal and Bateman, op. cit. note 1
22 Soal and Bateman, op. cit note 1; Martin and Stribic, op. cit. note 11.
23 Brown, Broadcast as cited; note 3 and op. cit. note 2.
24 Oram, A.T., 'An Experiment with Random Numbers', *J. Soc. Psych. Res.*, 1954, **37**, 369.
25 Even if Oram's work had confirmed Brown's suggestion in every detail, this would not have proved in the least that parapsychological data may be accounted for by such spurious effects. This will become clearer in the subsequent arguments given in the text.
26 Richmond, N., 'Two Series of PK Tests for Paramecia', *J. Soc. Psych. Res.*, 1952, **36**, 577.
27 This point of view is primarily due to Dr Harrop.
28 I consider only significance tests based on critical ratios, since for other tests similar arguments can be given.
29 Soal and Bateman, op. cit. note 1.
30 Ibid.
31 Martin and Stribic, op. cit. notes 14.
32 Pratt, J.G., and Woodruff, J.L., 'Size of Stimulus Symbols in Extra-Sensory Perception', *J. Parapsychol.*, 1939, **3**, 121.
33 Soal and Bateman, op. cit. notes 1.
34 Humphrey, op. cit. note 11.
35 Schmeidler, op. cit.note 11; Soal and Bateman, op. cit.note 1
36 Humphrey, op. cit. note 11 Murphy, op. cit.note 14.
37 The writer is indebted to Dr R.H. Thouless, Dr S.G. Soal, Professor J.M. Stratton, F.R.S., and the editor and referees of this *Journal* for helpful criticisms.

21 On the possible roots of consciousness and variations on themes by Hebb

21.1 Preliminary remarks

A little over 4 decades ago, the distinguished psychologist D.O. Hebb wrote a stimulating article titled 'The Problem of Consciousness and Introspection' (Hebb, 1954). I believe that some of Hebb's remarkable insights still merit renewed examination. I shall try to do so, but in the process I shall also add some ideas and views of my own.

To start with one may ask what is consciousness? Is it, like matter, an entity that can be described in terms of suitable physical parameters, such as mass, volume and so forth? This, surely, does not seem to be the case, since the linguistic attributes which we use to describe conscious experience seem to be totally different from the linguistic means employed. to characterize physical systems and their properties. For instance qualia, which are described by such things as a 'red patch' or a 'perceived mosaic of colours', or the perceived (experienced) shape of a cube, have certain properties which seem to be irreducible to physical thing language. Granted, the red patch emits light of particular wave lengths, and the emitted light is a physical property. But what we experience is not the emitted light, but properties induced (usually) within us, namely within our cranium. We experience a representation of the light-emitting object within us. Indeed this representation is, I believe, created by appropriate material systems (see below). There are, of course, many people who believe that a human being is composed of a physical body and a non-physical soul, and that it is the soul which is the seat of consciousness and which represents, or generates, specific conscious experiences, such as being aware of a red patch or the conscious experience of a cube.

Indeed, the red patch does not represent its 'redness' by being physically red in the sense in which it is experienced by us as red. The red patch emits quanta. But these quanta too cannot be said to be red, but to represent redness in the form of quanta of a particular kind. And so it goes on, via the retinas and to the brain. The retinas have colour receptive elements (cones) but these themselves cannot be said to be coloured or to perceive colour consciously. What we can certainly say, or believe, is that the redness of a perceived red patch is *created by us and within us* and likewise the perceived cubic experience of the perceived cube. But what about the system within us that creates conscious experiences? Is this system in any way like our somatic system that is made of flesh and bones? As regards the latter it is hierarchically built up of the kind of matter which physicists, chemists, molecular biologists and other material scientists have discovered. For instance the human body consists of bones and all kinds of softer tissues. All of these are made up of cells. The cells, in turn, are composed. of molecules, macromolecules, ions and so forth. The macromolecules, in turn, are composed of molecules, and the molecules consist of atoms and ions. Atoms and ions consist of atomic nuclei and electrons, and atomic nuclei consist of two kinds of quarks with the quarks of a nucleus being held together by gluon fields. We may conclude that ordinary matter, such as our human bodies, is ultimately composed of quarks and electrons. These then form, according to contemporary science the elements or 'building stones' of ordinary matter, as known to scientists.

Since consciousness does not detectably consist of ordinary matter, we may ask whether consciousness consists of 'elements' of another kind of substance or 'matter'. For those who believe that consciousness is part of and produced by, a human (or perhaps also animal) immaterial soul, the answer is simple. They argue that if there are any elements of consciousness, then these elements must be immaterial. On the other hand, in ancient Greece there existed already philosophers and thinkers who believed that there could. be a material soul (cf. Rist, 1972; Long, 1974) whose changes of states could be epiphenomenally accompanied by specific conscious experiences, and that consciousness belongs to, or is part of, that material soul. By contrast, according to these materialistic Greek thinkers, consciousness is not part of the ordinary brain, which is composed of quarks and electrons. But what warrants the hypothesis of a material soul, i.e. a material carrier of consciousness other than the ordinary body? I shall now discuss this issue.

21.2 Where is the seat of consciousness?

There are various reasons for believing that consciousness is *normally* related to the human brain, as known to anatomists, neurophysiologists and other biologists. In fact, since antiquity it has been known that brain lesions can impair consciousness. But can one conclude from this that the brain is the 'seat' of consciousness?

Not at all! Suppose that, as the Greek materialist philosophers thought, the brain is linked to a material soul. Then damage to the brain could lead to impaired communication between brain and soul. Hence, if the material soul is the 'real' seat of consciousness, and not the brain, then reduced communication between damaged brain and intact soul could lead to reduced or even completely abolished consciousness. Nevertheless, if one assumes that there exists no soul, then impairment of the brain alone might account for impairment of consciousness, on the assumption that consciousness is no more than a function of the brain and not a function of the soul. Thus, the effects of lesions data are equally compatible with either the assumption of the existence of a material soul or with the non-existence of such a soul and the existence of a brain only.

Next, let us look at the effects of electrostimulation of the human cerebral cortex, as performed by Penfield and Rasmussen (cf. 1950). When stimulated, some of the subjects would report that they experienced coloured sensations or in some cases, had 'dreams' which they reported. This, again, implies that brain activity is correlated with consciousness. But, again, it does not prove that the brain is the 'seat of consciousness'. It remains possible that a material soul, linked to the brain, is the 'real' seat of consciousness and that electrostimulation of the brain leads to activation of the soul. Activation of the latter could then result in the experienced colour sensations or the experienced dreams. It follows that, as with the brain lesions, the data cited up to now, do not rule out a material soul, but they also do not show that the brain alone is the seat of consciousness. I conclude that the preceding data, discussed up to now, do not show that the brain alone could not produce consciousness. Nor do they show that a material soul is necessary in order to produce consciousness. Thus, some of the 'classical' observations and experiments neither necessitate a soul nor do they rule out one. I am pretty certain that future manipulations of human brains, by methods not hitherto used, would lead to the same conclusions.

There exists, however, some evidence which, at least in my opinion, leads to different conclusions concerning the possible existence of a material soul. Before I proceed to this important alternative, which I developed first in an earlier book *'Shadow Matter and Psychic Phenomena'* (Wassermann, 1993)and which I had preceded in shortened form in the journal *Inquiry* (Wassermann, 1988), I must make here a short aside. The material soul (soon to be replaced by the 'Shadow matter brain' - see below) must not be confused with Descartes immaterial soul. He believed that there exists a dualism, in which an immaterial soul interacts, reciprocally, with the material body. In the present, wholly materialistic, theory the soul, though interacting with the body, is as material as the body, although, as we shall see, the matter which composes the soul is not made up of the same types of elements as the matter that composes our bodies. The matter which composes the material soul has been named *'Shadow Matter'* and is built up of shadow material elements.

At this stage the reader might wish to know what prompted my hypothesis of a material soul, or, as I shall call it, henceforth, a *shadow matter brain.* The basic observations which led to this are 'out of the body experiences', called OBEs for short. In a typical OBE the experiencer is aware of being himself outside his own body, and he may look upon his own body from the outside and recognize it as his own body. I have cited and interpreted several such OBEs in my book (Wassermann, 1993), and many more case histories and alternative interpretations can be found in the books of Blackmore (1982) and Green (1968). Let me just cite one typical case history of an OBE from Broughton (1991, p.244). He cites 'the experience of a British woman who was hospitalized and operated on for peritonitis. She subsequently became very ill with pneumonia and was confined to bed. The ward was L-shaped, and from her bed she could not see around the corner.'

> One morning I felt myself floating upwards and found I was looking down on the rest of the patients. I could see myself propped up against pillows, very white and ill. I saw the sister and nurse rush to my bed with oxygen. Then everything went blank. The next I remember was opening my eyes to see the sister bending over me. I told her what had happened; but at first she thought I was rambling. Then I said, 'There is a big woman sitting up in bed with her head wrapped in bandages; and she is knitting something with blue wool. She has a very red face.' This certainly shook her, as apparently the lady concerned had a mastoid operation and was just as I described. She was not allowed out of bed; and of course I hadn't been up at all. After several other details, such as the time by the clock on the wall (which had broken down) I convinced her that at least something strange had happened to me.

One way in which OBEs have been 'explained' is by assuming that they are hallucinations. This, however, is not plausible since many, or most, OBEs have typical class characteristics. It seems unlikely that in every OBE the experiencers just hallucinate the typical class characteristics when it is possible for them to hallucinate myriads of other experiences which have nothing to do with their OBEs.

A short time ago, I suggested a very different interpretation of OBEs. This followed a somewhat tortuous path. In the 1980s the theoretical physicist M.B. Green (now professor at Cambridge University, then at Queen Mary College, University of London) and others suggested that the so-called elementary particles of matter, such as electrons and quarks, were not, as had until then been generally believed 'point particles', whose mass was supposedly concentrated at a single point. Instead Green postulated that the mass of so-called 'point particles' was located in minute strings, so that, for instance, an electron was not located at a point but consisted of an immensely small string which incorporated its mass (see review by Green, 1985). This string theory became known as 'superstring theory' and was formulated as a higher dimensional theory. In 1985 Kolb *et al.* (1985) deduced from superstring theory in the journal *Nature* that there could exist another kind of matter, which differs from ordinary matter of, say, the kind of which human beings are made and which (ordinary matter) consists of quarks and electrons. Kolb *et al.* termed this new type of matter 'shadow matter'. Subsequently I suggested that apart from the relatively heavy shadow matter of Kolb *et al.* there could also exist an immensely light kind of shadow matter and that every quark could (reversibly) bind a shadow matter quark (made of this very light matter) and every electron could. (reversibly) bind a shadow matter electron (made of this very light matter) (Wassermann, 1993). It can then be argued that all ordinary matter objects, including human bodies, can bind precisely matching 'shadow matter objects' composed of shadow matter.

In particular, I assumed that to the human brain there is bound a matching shadow matter brain (composed not of ordinary matter but of shadow matter). Following this I assumed that the carrier of consciousness is not the ordinary matter brain, but the shadow matter brain that is bound to the ordinary matter brain. I assumed also that in an 'out of the body experience' (OBE) the shadow matter body (including its shadow matter brain) which is normally bound to the ordinary matter body, becomes detached from the ordinary matter body and floats away, subject to certain constraints which I have discussed elsewhere (see Wassermann, 1993). During the OBE it is the shadow matter brain, which via the shadow matter eyes of the shadow matter body performs visual perception. Similar remarks could apply to auditory perception. But it must be stressed that the shadow matter brain is in no way a duplicate, and certainly not an exact duplicate, of the ordinary matter brain, because the shadow matter brain is made, by hypothesis, of immensely light shadow matter and because this light shadow matter differs from ordinary matter in kind. (e.g. any shadow matter quark of the shadow matter brain differs in kind from the ordinary matter quark to which that shadow matter quark is bound. Likewise any shadow matter electron of the shadow matter brain differs in kind from the ordinary matter electron to which it is bound. The same remarks apply *mutatis mutandis* to the shadow matter quarks and shadow matter electrons of the rest of the shadow matter body which is bound to the rest of the ordinary matter body.) Thus in ordinary, everyday perception, thinking, feeling etc. it is, according to the Wassermann (1993) theory, the shadow matter brain, and not the ordinary matter brain which performs these cognitive functions. Likewise, in an OBE it is the detached shadow matter brain and not the ordinary matter brain which carries out all cognitive activities.

The immediately preceding remarks are important, since one particularly confused book reviewer thought that the shadow matter brain is assumed to be an exact duplicate of the ordinary matter brain, which, of course, is quite explicitly assumed not to be the case here and in my earlier book (Wassermann 1993).

The model put forward here provides at least one definite answer to the question 'where is the seat of consciousness?' It suggest that the seat of consciousness, both in ordinary cognitive activities and during OBEs, is the postulated shadow matter brain. I must also stress that my model is not based on superstring theory as someone who read Wassermann (1993) mistakenly thought. Although the notion of shadow matter (in Kolb *et al*'s (1985) version) was, indirectly, suggested by superstring theory, the particular notion of shadow matter suggested by Wassermann (1993) and in this chapter, makes assumptions about shadow matter which differ drastically from the shadow matter suggested by Kolb *et al*, as was made clear in Wassermann (1993). For instance masses of shadow matter quarks (= squarks) and shadow matter electrons (= selectrons) of the model discussed here differ drastically from the masses of the correspondingly named particles of the theory of Kolb *et al.* (1985).

According to the preceding model during an OBE the seat of consciousness (namely the postulated shadow matter brain) shifts away from the ordinary body and resides, temporarily, outside the ordinary matter body. Above all my model suggests that shadow matter brain activity, but not ordinary matter brain activity is essential for consciousness.

21.3 How is the 'seat of consciousness' genetically determined?

An important question is 'how could the genetic determination of the ordinary matter brain lead to a genetic specification of the shadow matter brain?' If it can be shown, or suggested, that the machinery that leads to the morphogenesis of the ordinary matter brain also leads to the formation of the shadow matter brain, then this could help to answer the preceding question. More than half a century ago the eminent embryologist Nobel Laureate H. Spemann (1938) noted on the final page of his Silliman lectures that, in his opinion, morphogenesis, say of the ordinary matter brain, and psychological processes are likely to be based on similar principles. Others have shared this view. So what are these principles? Although I have set out my conjectures repeatedly (Wassermann, 1972, 1978, Clowes and Wassermann, 1984) it is relevant to do so again, perhaps in a somewhat updated form.

In the 1930s the subsequent Nobel Laureate Professor Roger Sperry, of the California Institute of Technology, told Professor Paul A. Weiss that he believed that many, perhaps most or all, somatic cells were specifically and in many cases uniquely labelled. Weiss (1947) thought that the labels might be proteins. Obviously, not all somatic cells are uniquely labelled. Indeed, this is extremely unlikely for various types of blood cells. Yet, the tissue-specific cells of a kidney, or a lung, or a bone, could be each uniquely labelled. In this way mirror-image cells of a right and left kidney, which are in type corresponding, could differ with respect to many of their cell labels, and *mutatis mutandis* for the cells of many other organs or structures.

When Sperry first advanced the cell labelling hypothesis some people thought that each cell which is uniquely labelled carries a single type of unique protein label. But if there exists only one cell labelling protein type per uniquely labelled cell then this would lead to serious difficulties. A human brain could contain, say, 10^{11} or perhaps even 10^{12} uniquely specified neurons. If each of these neurons made just one protein type unique to itself and if each of these 10^{11} or more protein types were coded for by a single gene, then this would require 10^{11} or more distinct genes. This exceeds the estimated number of 10^5 genes of the human genome by several orders of magnitude. To escape from this dilemma I suggested (Wassermann, 1978) that cell labelling should be *combinatorial* with several cell labels per cell, an idea which, as far as I know, was neither envisaged nor worked out in terms of a model by Sperry. In fact, the combinatorial labelling of vast numbers of cells requires quite intricate combinatorial mathematics, and the underlying possible molecular biology is also very complex. A possible model was suggested by Wassermann (1978) and a formally improved version was presented by Clowes and Wassermann (1984).

The idea was that by means of suggested mechanisms in each neuron of the brain a *unique combination* of genes could be switched on, such that each of these genes could give rise to a unique type of protein. Accordingly, each neuron of the brain, according to this model, makes a combination of proteins unique for that neuron. Of course, different neurons could make some of the same types of cell-specific marker proteins (=CSMPs=cell labels), as long as the total combination of CSMP types per cell is unique. Similar states of affairs could apply to other organs (e.g. kidneys, liver, lungs etc.) and structures (e.g. bones).

It is now suggested that the particular combination of CSMPs of a cell generates a matching combination of shadow matter CSMPs which get bound to the ordinary matter CSMPs of that cell. In this way, and with the aid of additional induction of further shadow matter components, the genetically determined ordinary matter brain can bind shadow matter components which build up an indirectly genetically determined shadow matter brain which accurately matches (but not duplicates) the ordinary matter brain. (The matching between a shadow matter brain and an ordinary matter brain is in some ways, analogous to the matching of a photo and its negative.)

I have now discussed, in outline, how the ordinary matter brain could generate a shadow matter brain, which normally binds to the ordinary matter brain, although in an 'out of the body experience' (OBE)[1] the shadow matter brain, together with the rest of the shadow matter body, could become detached from the ordinary matter brain (and body) respectively. It is assumed that when the shadow matter brain (which might be identified with the material soul) leaves the ordinary matter body then *the shadow matter brain, as before, remains the seat of consciousness*. Also, the CSMPs are assumed to be the instruments of morphogenesis (cf. Clowes and Wassermann, 1984) as well as the generators of the shadow matter body, including the shadow matter brain. Accordingly the CSMPs are the providers of morphogenesis as well as the providers of a part the seat of consciousness.

I have argued elsewhere that the very light (or nearly massless) shadow matter which I am postulating as the ingredients of the shadow matter brain is likely to be a *very elastic* substance (Wassermann, 1993) and I discussed data consistent with this, or pointing to this.

The idea that a shadow matter brain coupled to the ordinary matter brain (or on its own in an OBE) could be the seat of consciousness, together with other explanations of psi-phenomena given in terms of shadow matter in my earlier book (Wassermann, 1993) aroused some interest. Thus, Nobel Laureate Professor Brian D. Josephson (Cambridge), a Fellow of the Royal Society, wrote to me after inspecting my book (Wassermann, 1993)

> The idea that some non-ordinary form of matter can explain psi is one I think may be worth following up, as you have done, in an attempt to get some way of thinking about these elusive phenomena. Purely as phenomenology your ideas may be very valuable.

Again one of the world's leading parapsychologists, the Carlson Professor of Psychiatry, Ian Stevenson, M.D., of the Division of Personality Studies, Health Sciences Center, University of Virginia, Charlottesville, VA 22908, wrote at the end of the review of Wassermann (1993) in *Journal of Scientific Exploration:*

> I heartily recommend this book to all investigators of paranormal phenomena and also to other scientists who would like to read a carefully written summary of the evidence for such phenomena. The book has adequate references and an index.

21.4 Psychophysical isomorphism

There have been many attempts to explain, in terms of theoretical models, how brains could produce psychological phenomena. I have surveyed and criticized a sizeable sample of these models and found them all wanting (Wassermann, 1978). This applies equally to my own model (Wassermann, 1978), although my book may prove helpful to future 'brain model' designers. One of the models which I always found most interesting and intriguing is Köhler's model (Köhler and Wallach, 1944). Köhler, an eminent gestaltist, thought that to shapes which are consciously perceived there exist similar shapes represented by appropriate structures in the human brain, notably on the cerebral cortex. When Köhler's 'psychophysical isomorphism' came under severe criticisms (cf. Wassermann, 1978, or Lashley *et al.* 1951) the idea of psychophysical isomorphism was abandoned. More 'modern' brain modellers thought that the human brain acts like a computer. Yet such models rarely, if ever, explain in coherent form the many findings of the Gestalt psychologists, like Köhler and others. The failure of extant brain models could be rooted in the theoretical choice of hardware on which these models are based. They are, without exception, hitherto based on known or postulated hardware of ordinary matter brains. Yet if, as I have argued (on the basis of OBEs) there also exist shadow matter brains in addition to ordinary matter brains with the former being normally bound to the latter, then, if the shadow matter brains are the seats of consciousness, it seems likely that shadow matter and its known, or postulated, properties may form the appropriate hardware for formulating brain models. In particular, the high degree of elasticity of shadow matter previously postulated (Wassermann, 1993, p.45, 46, 47, 48, 68) could permit important deformations of the shadow matter brain, and such deformations could form the basis of shadow matter explanations of Gestal phenomena. If so, then no amount of theorizing based on ordinary matter brains could hope to explain in a coherent and consistent form Gestalt phenomena and much else. Explanations of such phenomena could properly belong exclusively to the shadow matter brain. This, however, does not mean that the ordinary matter brain is not also involved in cognitive transactions. In fact, I have argued elsewhere (Wassermann, 1993) that the ordinary matter peripheral and central nervous systems play an important role in mediating information between the environment and the shadow matter brain.

In Wassermann (1978, p.2) I wrote that

> It is an elementary fact that the visual field is being mapped pictorially onto the retina, and that normal people report that subjectively they form three-dimensional (3D) pictorial representations of the visual world. The fact that a *pictorial* input mapping leads to a *pictorial* subjective experience, which

in special cases may very closely correspond to the input mapping, led W. Köhler to suggest that the brain representations which correspond to *pictorial* subjective experiences have themselves pictorial properties, i.e. that they form a *pictorial* brain mapping of the environment.

If, as Köhler did not suggest, and as is suggested in this chapter and in Wassermann (1993), the shadow matter brain, and not the ordinary matter brain, is the seat of conscious (i.e. subjective) experiences and of cognitive transactions, then it is the shadow matter brain representations which could correspond to *pictorial* subjective experiences, i.e. it is the shadow matter brain representations that form *pictorial* shadow matter brain mappings of the environment. This, of course, differs drastically from my earlier brain model (Wassermann, 1978) where it was assumed that the ordinary matter brain does all the mappings, and where shadow matter brains were not thought of (since I did not know of OBEs). Thus, Köhler's psychophysical isomorphism and his Gestalt theory surface here again, not in relation to ordinary matter brains but in relation to shadow matter brains. If so, then some, or possibly most or all, difficulties which confronted Köhler's original model, based on ordinary matter brains, might disappear. Let us take one example. Let us consider the mappings, distortions from the periphery on to the cerebral cortex. In this context I noted (Wassermann, 1978, p. 215) that 'In its cortical projection the area of the macular field of primates is magnified nearly 150 times, in relation to the perimacular region. Nevertheless, the perceived spatial extent corresponds to the extent of the sensory surface rather than that of the cortical area.' I added, 'Other very extensive mapping distortions, too numerous to relate here, are cited by Lashley *et al.*. (1951).' Thus, what has gone seriously wrong is the neural mapping from the periphery on to the cerebral cortex of the ordinary matter brain, so that the mapping on to the ordinary matter brain cerebral cortex does not correspond to the perceived spatial extent. If, however, the genuine mapping of the sensory surface which generates the conscious experience without mapping distortions is on to the shadow matter brain rather than the ordinary matter brain, then the difficulty of Köhler's model, for this particular item (and perhaps many others) could disappear.

The view that consciousness is the product of the ordinary matter brain is so widespread that it must come as a shock that things could be otherwise.

Before I conclude this section, I must write a few words about the relevance to consciousness of the now familiar 'split brain' observations by Sperry and others. Blakemore (1977, p. 159) while discussing the effects of sectioning the corpus callosum in human patients, cited Sperry's remark at a Pontificial Academy of Sciences conference, entitled 'Brain and Conscious Experience'. Sperry remarked that 'Everything we have seen so far indicates that the surgery has left people with two separate minds, that is two separate spheres of consciousness. What is experienced in the right hemisphere seems to be entirely outside the realm of awareness of the left.' These findings, however, do not imply that it is the ordinary matter brain and not the shadow matter brain which is the seat of consciousness. They only show that certain ordinary matter brain connections (i.e. the corpus callosum) have to be intact in order for the ordinary matter brain to act in an integrated fashion. All it means is that if the hemispheres are split apart, and if, say, the right hemisphere does not receive input from the left, then, according to the present point of view, the right hemisphere does not activate its attached shadow matter brain part, because of non-existent input. And, *vice versa* for the left hemisphere in case it does not receive any input. The split brain results are therefore consistent with the hypothesis that the shadow matter brain is attached to the ordinary matter brain and is the seat of consciousness. But Sperry's experiments and observations do not tell us anything new about the nature of consciousness (e.g. its 'seat').

21.5 Reality and manufactured consciousness

Suppose that we look at a green leaf. The leaf *per se* is not green, rather it is colourless. This means that the leaf can invoke colour in its beholder, but it cannot, really, except in everyday talk, be said to be colourless. Admittedly, when exposed to daylight, the leaf may give off light quanta. But, again, it cannot be said that these quanta are green. Also, when these 'green'-light conveying quanta strike a human retina, they may activate suitable retinal photoreceptors. But, again, these photoreceptors, as such, are not green. This process can go on until specific, postulated, shadow matter brain components are activated, whose activity then is epiphenomenally accompanied by the conscious experience of greenness. What applies to the particular example, just cited, applies *mutatis mutandis* to myriads of other examples. In other words, according to my theory, it is the shadow matter brain which, in response to a chain of environmentally initiated stimulus configuration becomes triggered in an internal response pattern which has conscious concomitants. Thus, the 'real' world is not the internally generated (or activated) world of shadow matter brain components which give rise to the conscious world of appearances. The real world is the external world which lacks the conscious properties of the hypothesized shadow matter brain, but which can, by the pathways indicated, stimulate the shadow matter brain and thus give rise to conscious experiences. (There were, of course, and perhaps still are, antiquated linguistic philosophers, who say that green leaves are green because we say so, and hence, they are green because that is how we use our language.)

21.6 Hebb's view of consciousness

Hebb (1954, p. 404) wrote

> I propose...that the existence of something called consciousness is a venerable *hypothesis* not a datum, not directly observable, but an inference from other facts.

This, however, sophisticated as it is, may be questioned. If consciousness is an epiphenomenon of the shadow matter brain, then an epiphenomenon is not a hypothesis, but could be a fact. The assumption that consciousness is an epiphenomenon is a hypothesis. It seems that Hebb has here substituted a hypothesis for a fact. We are asking here what sort of thing consciousness is and are answering that it is a hypothesized epiphenomenon. Yet, we know as little about what sort of thing an epiphenomenon is as we know what sort of thing consciousness is. Consciousness is not a thing in the sense that cars and chairs are things. If it is an epiphenomenon, then it could, indeed hardly be expected that an epiphenomenon is a thing in the sense in which ordinary matter objects are things. Most or all material things can, according to physics, be allocated physical 'dimensions' such as mass, length and time. This, however, does not apply to consciousness. We cannot ask how much mass a perceived red patch has. Perhaps this means simply that we have at least two kinds of entities in the universe, which have to be differently classified and which require, at least partly, different descriptions, namely physical objects and conscious entities such as qualia. We are concerned here with an irreducibility. The epiphenomena differ irrediucibly in kind from the physical things of which they are epiphenomena.

Does this mean the end of reductionism as known in the physical (including biological) sciences? Does it open the door to those fools who insist that 'reductionism is dead'? It does nothing of the sort. Reductionism is firmly established in the physical sciences and has already scored great triumphs and is likely to score further successes. What it means is that the ultimate entities in nature, say quarks, electrons muons, tau particles, and perhaps elements of shadow matter yet to be discovered, as well as consciousness, are not all entities of the same kind. Reductionism means that, for instance, a hydrogen atom's properties can be explained reductively in terms of its postulated constitutive electron and the quarks of its proton. Reductionism, however, need not mean that all properties of matter are, or must be, entirely explicable in terms of certain elementary physical entities.

If consciousness is not a hypothesis, but at best a hypothesized non-physical entity, then what is consciousness? Hebb (1954, p. 405) comes up with another suggestion. 'Assume then that consciousness is something within the head which determines behaviour and which can be studied through that behaviour.' *Pace* Hebb, if consciousness is a by-product (epiphenomenon) of behaviour-producing intracranial processes then consciousness does not produce behaviour. If so, then no amount of study of behaviour can reveal what sort of thing consciousness is, particularly if consciousness is a non-physical entity. For instance, consciousness could belong to the realm of 'shadow matter' (Kolb *et al*, 1985 or Wassermann, 1993), and shadow matter is a kind of matter which, by hypothesis, differs drastically from ordinary matter. Indeed, if there is a material soul, then it could be composed of shadow matter, so that a human being could exist of two mutually irreducible kinds of matter, namely ordinary matter (which is the subject of physics) and shadow matter (which might one day also form part of physics, but which might not be reducible to physical matter.)

Hebb, perhaps rightly, asserted that 'the least disputed criterion of consciousness is in certain verbal behaviour (introspective report) but this is not essential for we have no real doubt that a deaf-mute is conscious, even without sign language.' The trouble seems, to me, that Hebb was a dominant figure at a time when 'behaviourism' was the leading philosophy of psychology. It was then regarded to be disgraceful for a psychologist to proclaim that there could be anything that was not reducible to behaviour. This also comes out, at least indirectly, in Hebb's (1954) essay. He tries to lean consciousness on a 'certain verbal behaviour (introspective report)'. In those days it was simply not scientific, for ultimate heirs of Watson, to talk about introspection as something non-physical, so it was replaced by an 'introspective report'. Again Hebb (1954, p. 406) emphasizes that 'the higher animal is less stimulus-bound (Goldstein, 1940)...'

Hebb has rightly stressed, repeatedly, directly or indirectly, that consciousness has to do with ideation, or vice versa. Thus he writes (Hebb, 1954, p. 407) 'Consciousness in the higher organism constantly involves ideas about the future.' Hebb also stressed the significance of Lashley's (1951) important article on the serial ordering of thought and behaviour, which I read with great relish when it first appeared. The mechanisms of serial ordering are primary mechanisms of behaviour. Yet, I believe that we cannot say that consciousness, although an epiphenomenon, is causally produced by physical mechanisms of the kinds that produce publicly observable behaviour.

21.7 Sensory deprivation, hallucinations and consciousness

Elsewhere in his article Hebb (1954) describes the results of a sensory deprivation experiment. It was conducted by W. Heron, W.H. Bexton and T.H. Scott at McGill University. Hebb writes (1954, p. 415)

> The subjects are male college students paid to stay in a quiet cubicle on a comfortable bed 24 hours a day. They wear goggles which prevent pattern vision but admit light, and cardboard cuffs extending beyond the fingertips which permit free joint movement but little tactual perception... We have been able to induce our subjects to endure the boredom produced by these conditions for periods of two to six days.

Hebb notes that in these experiments 'the most striking result...is the occurrence of hallucinatory activity, chiefly visual but also auditory and somesthetic in a few subjects.'

Visual hallucinations, presumably, bear, as regards pattern organization, a resemblance to conscious visual perception. This suggests that if consciousness is a (shadow matter) brain-produced epiphenomenon, then this is produced similarly in normal consciousness and in the hallucinations under discussion.

In the discussion, following Hebb's (1954) paper, there is an interesting remark by A. Fessard, who asserts

> I do not quite agree with Dr. Hebb's definition of consciousness in terms of behaviour. I am quite sure I can accomplish complicated mental operations with the same behavioural manifestations as Dr. Hebb has associated with consciousness but without being aware of their being performed. At least a special mention should be made of a kind of behaviour that is characterized by its power of evoking more or less similar states of consciousness in minds of others, that is *language*. Verbal reports by the subject of his having just been aware of some percept, memory or any other mental operation of his own, seems to me the really useful criterion for consciousness, if at times unsafe.

Fessard's remarks illuminate the complexity of the problem. There is seemingly an encoding of consciousness into language. But this may be deceptive. The physical (shadow matter) brain structures whose epiphenomenal concomitants could be particular conscious experiences could give rise to the encoding into language, so that it is not consciousness itself, but rather the non-conscious physical partner system of conscious experience that becomes encoded into the brain system that communicates language. That people can communicate by speech is due to the, assumed, circumstance that the brain of the language user, as well as that of the language perceiver both construct and store in physical form what is to be spoken and what has been perceived (via language). That these constructive physical systems, when activated, could both have conscious concomitants (i.e. the language producing system of the speaker and the language perceiving system of the perceiver, or else systems associated with these systems) may give us the deceptive impression that consciousness of the speaker communicates with consciousness of the perceiver. But this may simply not be so, according to epiphenomenalism.

Finally I should like to consider the last comment by Hebb (1954, p. 421) on consciousness. He claims that 'consciousness is a complex inference from what movement is eventually observed of the speech organs or otherwise'. I am not sure that nowadays, in 1995, many people would subscribe to such a radical behaviourist criterion of consciousness. Suppose we went to another planet and met people that looked rather human and could speak, but who anatomically differed from us, particularly as regards brain structure. Could we be sure that they have conscious states? I believe that we can only be sure that other people are conscious in the way that we are conscious provided they have brains which are closely similar to ours and provided what they say invokes in us conscious states that 'make sense'.

21.8 A possible major implication

In Section 21.2 I drew attention to 'out of the body experiences' (OBEs) which have been reported by numerous people, and which, notably, may occur as components of 'near death experiences' (NDEs), although not all NDEs exhibit OBEs. I assumed that in an OBE the shadow matter body becomes detached from the ordinary matter body and can float some distance away from the ordinary matter body. There remains the important possibility that when death occurs the shadow matter body (including its shadow matter brain) separates permanently from the ordinary matter body. This could lead to temporary, or possibly permanent, survival of the shadow matter brain, and all its memories and cognitive faculties, after death. If so, then OBEs and NDEs provide the best possible indirect evidence for the survival of the human personality after death (cf. Wassermann, 1993). This brings us back to the long-existing view, favoured by many religious people of various religions and various civilizations, that the soul (here identified with the shadow matter brain) can survive, possibly permanently. The difference, however, between the present view and purely religious belief is that the possibility of survival of a material soul, and the existence of such a soul (bound normally to the human body during its life) is here based on experienced

OBEs (including the OBEs of NDEs). In fact, some hypersceptics, who put their beliefs above the facts, will obstinately maintain, in the face of a wealth of evidence to the contrary, that OBEs do not exist. Perhaps such people ought to see a psychiatrist. Such people firmly believe that survival of a material soul is absolutely impossible and that every one who does not accept their views is a fool. Yet, contrary to their dogmatism are the views of those who accept OBEs and their possible implications of survival.

Finally there are writers who either repudiate the existence of OBEs or, who, like Greenfield (1995), disregard them. Such writers adhere to the traditional view that, say, consciousness is related to the ordinary matter brain. A slight modification of Greenfield's (1995, p. 9) view reads thus:

> One can very easily imagine that individual brains, or rather their intrinsic neuronal connections, are gradually shaped [developmentally] or by experience, and that in turn these modifications to the brain are constantly modifying the mind and hence the consciousness of an individual throughout life.

The fact, however, that the human mind in most cases grows as the brain grows, does not imply that the mind, and its consciousness is necessarily directly controlled by the ordinary matter brain. The existing facts (and some not mentioned by Greenfield, e.g. OBEs) are consistent with the hypothesis that the human mind and its consciousness are linked to a shadow matter brain which is linked to the ordinary matter brain (as suggested in detail in Wassermann, 1993), and that as the ordinary matter brain grows so does the shadow matter brain.

Note

1 In passing I must state that many OBEs occur as components of near death experiences (NDEs) (cf. Blackmore, 1993 for a discussion of NDEs).

Bibliography

Blackmore, S.J. (1982) *Beyond the Body*, London, Heinemann.

Blackmore, S.J. (1993) *Dying to Live*, London, Harper Collins Publishers.

Blakemore, C. (1977) *Mechanisms of the Mind*, Cambridge, Cambridge University Press.

Broughton, R. (1991), *Parapsychology: The Controversial Science*, London, Rider.

Clowes, J.S., and Wassermann, G.D. (1984) 'Genetic Control Theory of Developmental Events' *Bull. Math. Biol*, **46**, 785-825.
Fessard, A.E. (1954), 'Mechanisms of nervous integration and conscious experience', in *Brain Mechanisms and Consciousness*, J.F. Delafresnaye (ed.), Oxford, Blackwell, pp. 200-236.
Goldstein, K. (1940), *The Organism*, New York, American Books.
Green, C.E. (1968) *Out-of-the-Body Experiences*, London, Hamish Hamilton.
Green, M.B. (1985), 'Unification of forces and particles in superstring theories', *Nature* (London), **314**, 404-414.
Greenfield S.A. (1995) *Journey to the Centers of the Mind*, Towards a Science of Consciousness, New York, W. H. Freeman.
Hebb, D.O. (1954), 'The Problem of Consciousness and Introspection', In *Brain Mechanisms and Consciousness*, J.F. Delafresnaye (ed.), Oxford, Blackwell Scientific Publications, pp. 402-421.
Köhler, W., and Wallach, H. (1944), 'Figural after-effects: an investigation of visual processes' *Proc. Amer. Philosoph. Ass*, **88**, 269-357.
Kolb, E.W., Seckel, D. and Turner, M.S. (1985) 'The Shadow World of Superstring Theories' *Nature* (London), **314**, 415-419.
Lashley, K.S. (1951), 'The Problem of Serial Order in Behavior' in L.A. Jeffress (ed.) *Cerebral Mechanisms in Behavior*, New York, Wiley.
Lashley, K.S., Chow, K.L. and Semmes, J. (1951), 'An examination of the electrical field theory of cerebral integration', *Psychological Review*, **58**, 123-136.
Long, A.A. (1974) *Hellenistic Philosophy*, London, Duckworth.
Penfield, W., and Rasmussen, T. (1950), *The Cerebral Cortex of Man*, New York, Macmillan.
Rist, J.H. (1972), *Epicurus*, Cambridge, Cambridge University Press.
Spemann, H. (1938), *Embryonic Development and Induction* (Silliman Lectures), New Haven, Conn., Yale University Press.
Wassermann, G.D. (1972), *Molecular Control of Cell Differentiation and Morphogenesis*, New York, Marcel Dekker.
Wassermann, G.D. (1978), *Neurobiological Theory of Psychological Phenomena*, London, Macmillan.
Wassermann, G.D. (1988), 'On a Physical (Materialistic) Theory of Psi-phenomena Based on Shadow Matter', *Inquiry*, **31**, 217-222.
Wassermann, G.D. (1993), *Shadow Matter and Psychic Phenomena*, Oxford, Mandrake of Oxford.
Weiss, P.A. (1947) 'Problem of specificity in growth and development' *Yale J. Biol. Med*, **19**, 235-278.

Author Index

Subject Index